Border Bees

Colin Weightman, MBE

Northern Bee Books

Border Bees

A reprint of the 1961 edition, revised and edited by John Phipps, with an additional chapter on Heather Honey Production, and extracts from Colin Weightman's columns which appeared in the British Bee Journal.

Published by Northern Bee Books 2012

ISBN 978-1-908904-19-5

Northern Bee Books
Scout Bottom Farm
Mytholmroyd
Hebden Bridge HX7 5JS (UK)

Border Bees

Colin Weightman, MBE

To my Parents and Margaret

Northern Bee Books

CONTENTS

Foreword   John Phipps

Foreword to the First Edition - Mr J. S. Cox

Glossary

Preface

Chapter 1. Introduction

Chapter 2. Farming Forebears

Chapter 3. The Honey Makers

Chapter 4. Years Good and Bad

Chapter 5. High Summer

Chapter 6. Heather-Going

Chapter 7. A Winter's Tale

Chapter 8. Royal Jelly and All That

Chapter 9. Problems

Chapter 10. People

Chapter 11. A Complete Guide to Heather Honey Production

Chapter 12. Extracts from Colin Weighman's Columns in the
British Bee Journal:
*One hive four queens; Winter of 1962/1963; Winter
Wonderland; Wet Summer of 1985 Responsible for Winter
Losses; Swollen Streams Flood Apiary; The Heather Beetle;
Drones; Oil Seed Rape Honey; Rhododendrons and Bees;
Thefts of Queens and Colonies; Crown Boards, Floorboards
and Over-wintering Bees; Illingworth's Oilcloth Excluders;
Removing Hives From the Moors; Making Up Nucleus Colonies
in Northumberland; Country Life; Mountain Grey Apiaries;
A Visit to Mr Snelgrove; Donald Sims; William Hamilton;
Heather People: John and Rosemary Theobalds and the
Robson Family of Horncliffe; Visit to Eva Crane at Hill House;
In Greece with Brother Adam; St Ambrose of Milan, the Patron
Saint of Beekeeping.*

APPENDIX Comparative Costs of Hives and Bees (1959)

INTRODUCTION

'Some have alleged that a portion of the divine mind and
A heavenly emanation may be discovered in bees'

Virgil

Many books have been written about bees but this is one with a difference; a suitable read for novices, experienced beekeepers, those who may be thinking of taking up bees or those who do not keep bees at all. It may be considered controversial, perhaps even heretical by some, but it is written honestly and sincerely with the welfare of the bees in mind.

To really understand bees takes a lot of observant study with lots of bees over a long period of time; a lifetime in fact. You see, the active beekeeping season is very short in this country and the person with one or two colonies can gain very little experience in a single season or in ten seasons for that matter though that may be sufficient for his small-time needs.

In working with a hundred or more colonies he will begin to learn what the bees really have to offer in the way of teaching us. This is not a practical course of action for the average man or woman and therefore, if they are to embark on keeping bees they must get that practical experience by talking with other beekeepers, visiting their apiaries and learning from their triumphs and disasters or perhaps reading the odd text book.

I am the third generation beekeeper in my family and my daughter is becoming the fourth. Since early childhood I was involved in bees and have kept them large scale and small scale and I have had some wonderfully interesting times. I have also had my share of calamities and made many mistakes and this is the reason for writing this book, if only to save others from a few of those.

This is not a text-book on how to keep bees nor is it particularly scientific. There are plenty of those. Some are very good, one or two are excellent but it has to be said, for me, many of them seem to be written by enthusiastic amateurs with only a limited grasp of the real needs of the bees. My father did not say much but what he did say made sense. One of his quotes was, 'You can teach the bees nothing; they can teach you everything, so let them'!

He kept a stool set between his two hives and there he sat after a day's work and watched the bees, marvelling at their industry, intelligence and bravery so why not do the same!

Investiture

ACKNOWLEDGMENTS

Since 1950 Mr. and Mrs. Cecil Tonsley have encouraged the keeping of bees in Britain in their own particular way. The author records his appreciation.

FOREWORD TO
SECOND REVISED EDITION

I have, much to my disappointment, never met Colin Weightman. However, ever since I started beekeeping over forty years ago I have been aware of this great Northern beekeeping personality, whose columns I looked forward to reading each fortnight when I bought the British Bee Journal from my local newsagent. Over the years, through his writings, I learned much about the arduous work of the northern beekeepers, but more than that, for Colin framed it in context with his beloved surrounding countryside, which he described lyrically and with interesting anecdotes of his observations of nature, the weather and farming. Beekeeping is very much part of the living landscape, the craft cannot be looked upon in isolation, and fortunately we have in Colin a real rural character; beekeeper, farmer and nature lover who has also delved into his local history, to serve as a reliable and entertaining commentator on country life over eight decades.

Happy as he is in his native county, Colin has travelled extensively throughout England and Scotland and the many anecdotes recalled in 'The Border Bees' and the BBJ, give us a good picture of what many people believe to have been the golden years of beekeeping, before the coming of varroa, and the many characters who were prominent in the craft.

Colin's travels also took him abroad, sometimes accompanying Brother Adam, the famous bee breeder, who Colin met at an early age and with whom he kept contact until the end of the monk's days. Undoubtedly Colin was greatly influenced by this Buckfast beemaster, but his Northumberland roots and the ways of the heather men of yore had an equal hold on him.

Colin has written for us several times in The Beekeepers Quarterly, including a detailed and practical account of heather honey production, which we have added as an extra chapter to this book. Additionally included are extracts from Colin's columns written for the British Bee Journal over many years.

I am very honoured to have had the task of updating and editing this new edition of Border Bees. I hope that for older members of the craft it will bring back pleasing memories of times gone by and for newcomers some very welcome sound advice from an eminent and practical beekeeper.

John Phipps
Greece
June 2012

"When the summer's on the wane, and the corn begins to fall,
When the lime has drop't its blossoms and the honey flow is small
Then the beeman rises early for he hears the heather call,
To the fell with the bees in the morning".

There is colour on the moorland there is fragrance in the pine,
There's a breeze upon the mountains that intoxicates like wine,
The clouds have golden borders where the sun begins to shine
On the fells with the bees in the morning".

From: The Road to the Fells
By the late R E Richardson, Stocksfield, Northumberland.

FOREWORD TO THE FIRST EDITION

By Mr. S. J. Cox, NDB, Shropshire,

(Formerly head of the Horticultural Department,
Houghall Farm School, Durham)

To find a niche for a new book on bees is not easy, but TALES OF A BORDER BEEKEEPER has, I feel, done just that. It is rather unique, as it introduces the reader to characters, personalities and eminent beekeepers, some of whom, fifty years ago, inspired Colin when he took up the craft. Much of the book is anecdotal and, because of his travels, not nearly so parochial as the title might suggest. During the thirty years that I lived in neighbouring County Durham, I spent many a happy day with Colin in his apiary at Shilford, and it was there as much as anywhere, that I came to appreciate the joys of beekeeping. Then, there was the hospitality - a Weightman tea - in the cool farmhouse kitchen, was quite something! For long Colin has been a mentor and practical help to both new and experienced beekeepers, always ready with an idea or two. It always strikes me that he measures his return from beekeeping, to some considerable extent, in terms of friends: a nice balance. May he continue to make a contribution, for many years, in his inimitable manner.

S. J. Cox, Shropshire, March, 1961

GLOSSARY

1. BRAIRD - Growth.

2. EKES & COGS — Additional honey storage chambers used by the beekeeper at the heather. The bees are allowed to build comb naturally, or from starters of foundation.

3. FELLS — Moorland or hill country in the North of England.

4. FETTLE — Condition.

5. HAUGH LAND — Light sandy land found mostly in river valleys.

6. IN BYE — Precincts of the home or farm.

7. MAIDEN SWARM — A name given to a swarm which comes from a swarm from earlier in   the season.

8. SKEP — Old type of fixed comb straw hives. Empty skeps sometimes used for taking swarms.

9. STANCE — Apiary or stand for hives of bees at the heather.

PREFACE TO THE FIRST EDITION

"It is what we learn after we think we know it that counts ".

THERE have been few attempts to describe bee-keeping practices in the moorland areas of the British Isles.

I regard the present work as a "peace offering" to friends who have over the years asked for an account of my bee- keeping experiences. I have told the story against the back-ground of the Roman Wall—a part of England with a charm of its own.

Improved strains of honey bees, systems of colony management, with tolerant guidance from Education Authorities as well, place the craft of bee-keeping in the very forefront of rural pursuits today. For those who seek a complete form of relaxation, bee-keeping on the right lines can become one of the most absorbing and stimulating undertakings known.

I am indebted to the following friends for their scrutiny of my manuscript and for other incidental help: Professor H. Cecil Pawson, Emeritus Prof. Agriculture, Durham University;   Mr. F. Austin Hyde, formerly Headmaster Lady Lumley Grammar School, Pickering, and Woodhouse Grove, Yorks; Mr. L. E. Snelgrove, M.A.. M.Sc., Bleadon, Somerset; and Mr. George Jobson, Agricultural Editor, "Newcastle Chronicle and Journal." The imperfections are my own.

Colin Weightman
Shilford
Stocksfield
Northumberland
January 1st. 1961.

This historic photo (circu 1935)
Taken at Robson & Cressfords Heather Stance at Harwood Shield, Hexhamshire
Left to right J. Raine, George Kay. W. Arthur, Bob Robson, Frank Cessford - with pipe
front row Munro Sutherland, Hon. Sec Newcastle & District BKA.,
Annie Betts - Editor Bee World,
Dr. John Anderson (from Aberdeen) noted Scottish Beekeeper

CHAPTER ONE
INTRODUCTION

"Honey each night
Before repose,
Will slam the door
On the Doctor's nose".

IT was an unbearably hot day. Rocks, rowans, and an occasional pine, overhung the moorland stream. For once, the wind had dropped and a heat haze simmered across the valley. I stood on the heather moors of Northumberland in August. The scene showed the splendour of the purple heath. Removed, and in the background, there was the drowsy hum of bees as they toiled laboriously amongst the flowers of the ling. It was another heather time, and on moors and hills in many parts of the British Isles that group of people who followed the ancient craft of bee-keeping were doing a similar kind of thing. In misty imaginings as to how this had come about and how I had got caught up in it, I recalled my first experience of bees when I visited my Grandmother on her small farm in Northumberland in the 1930's, and helped on occasion in the garden, which was an unnecessarily large one, but, at the same time, meticulously kept. Gardens were a fetish with her, but it was there that I was brought face to face with honeybees—and I disliked them very much !!

The gardens, as I remember them, were attractively displayed and contained numerous long, narrow paths, flanked on either side by golden rod. This was aglow with colour in late July and August, but it attracted insects by the thousands, and bees into the bargain from George Kay's nearby hives (it was years later that I discovered that the owner of these bees was quite a celebrated figure in his calling and one of the best bee men the North has produced). But all this was unknown to me then, and I flatly refused to pass those bustling avenues when the sun was shining.

After farming extensively in the County Palatine of Durham for over one hundred years, my family returned to Northumberland in 1938 to escape industrial encroachment, and occupied a farm of free working haugh land in the Tyne basin.

Those who travelling to Scotland know the county from the train only, and the impact of Newcastle and industrial Tyneside, can know little of the Northumberland allurement. One of the great grazing counties of England, gently, undulating country, and umbrageous parkland, stretches to Tweedside and the Border. To the East, there is a coastline rich in wild life, while to the West lie fells which take in the Roman Wall and loughs (Northumbrian Lakes).

The magic of the Wall country must be seen to be believed. The strange grandeur of the crags as they fall abruptly into Crag and Greenlee loughs. Colourful little sailing craft, skimming under full sail across the Greenlee waters; the occupants actively engaged in racing. I have drifted lazily on the lough, the pools becalmed, and have seen its face lashed in the sudden fury of a squall. One is always conscious of the past and the fact that centuries earlier Roman Legionaries had looked Northwards upon a very similar scene. Bonnyrigg Hall, Gibbs Hill, and Sewingshields resound the character of the place and cast their own particular spell.

The Roman Wall - Here the Wall makes its splendid progress across the rolling Northumbrian uplands. "One is always conscious of the past and the fact that centuries earlier Roman Legionaries had looked northwards upon a very similar scene."

This is a countryside which lends itself to crafts; the home of individuals, unhurried in their ways and outlook, and endowed with qualities as enduring as the hills themselves. In short, a natural store house—or reserve—from which the country can draw new strength after the disillusion, listlessness and inertia of the World Wars.

The agricultural pattern set during the long reign of George III was to endure through both the world upheavals (1914-18, 1939-45), and, despite an increase in the migration of the rural population to the towns, those who tilled the land at home became machine-minded, and made full use of the great scientific and technical developments of their time.

My bee-keeping apprenticeship proper began during the first years of World

War II when as a schoolboy I met the late Jack Tweddell, of Riding Mill. This in turn led to our buying for the modest sum of shillings, a hive of bees from a very lively old lady, now dead, who lived on the outskirts of the village.

First Hive, 1940.

Colin Weightman, July 1947.

Riding Mill BKA visited me in a later year (1957) at Shilford.
Left to right: George Tweddle (Hon Sec), Bob Kerr (who followed him as Hon Sec),
Frank Cessford (seated), behind him Fison, Jack Tweddell
(the Riding Hill post master who brought the first colony to Low Shilford in 1940 and
introduced CW to beekeeping), M Grossick (Corbridge), Colin Weightman.

Even now I find it hard to believe that this chance meeting was destined to have such a far-reaching effect upon my life. If this friendship had not come about—and my interests until then lay in an entirely different direction—a whole tableau of events and way of life would have gone by unnoticed. An interest in Italian bees soon led to correspondence with Brother Adam, the famous bee-keeping monk of Buckfast Abbey, Devon, and many other notables in the bee-keeping world. With the end of hostilities it was my privilege to meet Brother Adam for the first time, and the story of this trip to Buckfast and the famous queen which I obtained from him has often been told against me in the North. Petrol rationing was, at this time, still vigorously enforced, so I decided to write in to the County Agricultural Officer to state my case. I emphasised the need for me to collect the queen in person.

The County Agricultural Officer was sympathetic, and back came the petrol coupons—30 gallons! Years afterwards. Mr. John Davies, former Principal of the Cumberland and Westmorland Farm School, Newton Rigg, one of the most dynamic figures in farming circles, visited us at Shilford to demonstrate at a Shorthorn Society Field Day, and he recalled how, as Assistant to the County Agricultural Officer, he had made this visit possible. I returned from Devon with an indelible impression of Brother Adam's life work; the culmination of years of effort and endeavour. I brought something even more tangible—the queen! For she turned out to be one of the best queens I have ever had. Brother Adam and I have kept in touch with each other over the years, and in 1955 the Prior of Buckfast, Dom. Leo Smith, spent a day with us at Shilford.

The Revd Brother Adam, Buckfast Abbey, at the door of
Home Apiary Grafting House, May 1948.

Northern Federation Meeting, 1948.

Newton Rigg, 1950.

On the return run from Buckfast Abbey I visited—also for the first time—another of Britain's foremost bee-keepers, Mr.Snelgrove, of Bleadon, Weston-Super-Mare, whose work in connection with swarm control will endure for all time. Years later we met at the British Bee Journal offices in London during Mr. Snelgrove's term of office as President of the British Beekeepers' Association.

I shall remember Snelgrove as a rather tall upright old gentleman, kindly, with a particularly well-modulated voice. He was a man who inspired confidence in young people. One of my most treasured beekeeping possessions is an inscribed copy of his book "Swarming—its Control and Prevention," which he presented to me in 1959.

Throughout this period, and in fact right up to the present time, I have had the inestimable advantage of a friendship with a true son of the soil—Old Bill, the Gamekeeper. Bill Hindmarsh has had a fifty years association with this part of Northumberland. A loyal and dedicated retainer to a family whose interests he had nurtured for so long, his is a classic example of family veneration which stands out like a beacon in an age of crumbling national values. It was he who accompanied me on many boyish escapades to watch the little owls and the badger colonies in the great earths of the pine woods near my home. The occasional invitation to take part in a ferreting expedition, and the opportunity to track wild life through the frost and snow of the winters of the 40's are very cherished memories. As my "Gamekeeper", old Bill has become well known in the "British Bee Journal" and in recent years has accompanied me all over the country, delighting all who have crossed his path.

August 1947 was renowned for its long spell of dry weather which extended over almost the whole of the British Isles. Perfect cloudless conditions continued day after day: and for once the cool east winds which we are accustomed to during these anti-cyclone periods were entirely absent. As you can imagine, the honey flowed…

Hodgson Gray and Gambles (1955).

The Tyneside Agricultural Society staged its annual one-day exhibition on the Bank Holiday Monday, and on this occasion the beekeepers of the area were invited to make a contribution. I got hold of a tent, and took a hive or two of bees along, just for the fun of the thing, and we had a most successful day. A steady stream of callers from all parts of the country—interested in bees—made themselves known. Towards evening the late Joe Jordon, well known for his Longhirst Shorthorns, came into the tent accompanied by a striking personality in a deerstalker hat, whom he introduced as Hodgson Gray. A friendship was forged there and then. This too has endured. In the field of beekeeping it is doubtful if anyone has had greater experience and success in so many areas. He has been a prolific writer to the bee press since the 20s: and he takes his stand by the brown bee indigenous to the North of England. Over the years he has developed his own distinct line. Mrs. Gray is a delightful personality in her own right. For most of World War II they were more or less entombed in the old shooting box, high up above Rothbury and Alwinton, known as Kidlandlee. To get there you traversed the most difficult of moorland tracks, crossing and recrossing runnels and small water-courses until you made the final ascent to the eyrie itself. A grim place, lost half the time in mist.

We turn now to farming forebears.

CHAPTER TWO
FARMING FOREBEARS

"We tread the paths their feet have worn.
We sit beneath their orchard trees,
We hear, like them, the hum of bees
And rustle of the bladed corn".
(WHITTIER — "Snowbound")

The farm at Lower Shilford. (Shilford Farm is situated midway between the villages of
Riding Mill and Stocksfield on the A695 Newcastle-Hexham road.
It is 14 miles from Newcastle and 7 miles from Hexham. Riding Mill lies on the main
A68 Scotch Corner/Jedborough road to Scotland).

Colin's parents, Winnie and Harry Weightman with family friend Alice Hunter., 1960.

THE majority of people consider this subject so musty—brevity will be the keynote.

My forebears were Normans [1]. They came to England with the Conqueror and subsequently held land in Leicestershire and the adjoining counties: being particularly associated with Burbage, The Manor of Wiken, and the old family of Trussell [2]. During the seventeenth century they moved North into Northumberland and Westmorland. The Northumberland family farmed "Overthwarts" and then came to Corbridge in the valley of the river Tyne, just under the Roman Wall, and from there repaired to Sunderland. From the time of the Napoleonic Wars they have been interested in potato growing, and this particular leaning has continued until the present time. As the period embraced both lean years and times of plenty, it became necessary to evolve a policy which was at once sound and which would withstand the test of time. Thus cash cropping, in the form of wheat and potato growing went hand in hand with dairying. At Herrington and Silksworth, near Sunderland, my great-grandfather and his brother brought large acreages under cultivation for the first time. At Herrington the carelessness of a horseman, when he had finished rolling a field of winter wheat, resulted in the roller taking in some recently sown drills of turnips as well. The year turned out to be an exceptionally dry one, and the rolled drills had a "braird" of turnips, which are reputed to have stood out a mile!!

By accident, one of the most important aspects in growing turnips successfully on light land—the rolling of the drills immediately following rowing up and sowing to conserve moisture—was hit upon for the first time. At once the news spread far and wide, and turnip growing everywhere under these conditions was made a little easier. It was at Herrington, in the early part of this century, that Albert Weightman pioneered the breed of British Friesian cattle in the Northern parts of England, and he was the first North countryman to become breed Society President. A man of great drive and foresight he visited Holland in the 1930s and through his influence the British Friesian Cattle Society made its historic importation of Dutch cattle in 1936. An event destined to change the whole face of British Dairy Farming practice and often hailed as one of the most revolutionary farming issues of the twentieth century. His own imported Dutch bull, Herrington Leo, became one of the country's best known animals, and together with some female lines, introduced butter-fat into the herd to such a point that it became well-nigh unassailable in the breed. The Herrington herd is continued today under the Warden prefix of my cousin Wesley and his family. One thing became apparent to my grandparents in those early days: cattle had to have bone and size, and plenty of it. After a useful milking life, the cows had to have a carcase which would please the butcher—and the same thing applies today.

By the end of World War II the Friesian had almost replaced the Shorthorn in this country as the poor man's cow. But both breeds are now acutely embarrassed by the loss of size. Breeders must give this matter thought and provide cows capable of producing quality milk in quantity—cows able to fill three important roles.

The Westmorland family were living at Appleby at the time of Waterloo. A son, William, came to Durham University in 1835. He eventually took Holy Orders, was ordained, and received some notoriety as suitor of Emily Bronte who wrote "Wuthering Heights" and was to become the most famous of the three literary sisters from Haworth Parsonage.

Another forebear left this country for the USA and founded his own branch of the family there. These American Weightmans were great beekeepers, and one of the girls, Helen, bought the "Californian Bee Journal" and was Editor for some years.

Colin Weightman continued with the family's interest in cattle breeding. Commonwealth marketing problems, concern over European Free Trade, and an invitation to discuss bee-keeping matters as well, led him to Denmark in the summer of 1960. The picture shows him amongst the Danish Red Milkers, a breed destined to play an increasingly important part as an improver of dairy types of cattle (with similar colour genes) in all corners of the globe.

My old friend, Dr. Malcolm Fraser, for years one of the world's most distinguished scholars on the historical aspects of beekeeping, indicates that the duties and lore of the bee herd were passed from father to son over hundreds of years, and, as a leaning towards bees crops up again and again in succeeding generations of my own family, this provides interesting speculation.

(1) F Anson's "History of the Weightman Family"

(2) Burke's Landed Gentry (4)

CHAPTER THREE
THE HONEY MAKERS

"Where bees are there is honey".

PICTURE a large unspoilt area of Britain, a sweeping coastline, rolling hills just touched with purple, air that has a tang in it. You will then come very close to knowing and understanding the Northumberland of this chapter.

The four counties which make up Northern England are divided by the chain of Pennine Hills. Rainfall is heaviest to the West, and in the Lakeland valleys differences ranging from fifty inches to one-hundred-and-fifty inches occur within a mile or two. In direct contrast to the West, the Eastern coastal areas of Northumberland have an average rainfall of under thirty inches: here clover flourishes on an expanse of carboniferous limestone; and these, as you can imagine, are honey-producing areas of importance. The river valleys, where *haugh* land predominates, and the scarpland with its shale, provide only secondary flows.

We attach great importance to the fact that the ling heather *(Calluna vulgaris)*, from which our main honey crop is gathered, is found only in profusion on Eastern slopes. It is noticeably absent in the West owing to the richer lime deposits and heavier rainfall. As a general rule, rainfall must not exceed thirty-five to forty-five inches to permit definite soil drying in summer; it is only under such conditions that heather will succeed.

In Northumberland's case marginal farmland, forest, and fell divide much of the agricultural land into small areas, with the result that beekeeping is restricted to the Tyne basin, the rich pasture land of the coast, or this side of the river Tweed, celebrated for its salmon. This part of the North-East has a charm all its own.

Over a large part of the area beekeeping follows traditional lines introduced with the movable comb hive at the end of the last century. Brown bees predominate, and 9/10 British Standard combs are given to each colony to develop to its full capacity. With the presence of large industrial towns, and the present full employment, there is an excellent demand for honey. Sections are a most attractive proposition, and seasonal work consists of encouraging and taking early natural swarms. These replenish the empty hives, which are present in almost every apiary after a severe winter. A handful of what we might term undaunted beekeepers, work with more prolific strains and follow a long-term policy. This necessitates providing extra breeding room for the colony, by allowing the queen access to a brood chamber, and a super, or alternatively, two standard brood chambers.

Sections are always an attractive proposition…

I suppose I offend the popular belief that a dark bee is best suited for the Northern parts of England. This belief is not substantiated in practice [1]; and I take my stand with a yellow-banded bee—particularly the line developed by Brother Adam. In appearance it is similar to the Italian, but it is fortunately endowed with many important characteristics the popular imported strains do not possess. I consider this yellow-banded bee superior in many ways to the local bee. Brother Adam's yellow-banded bee requires a brood chamber equivalent to twenty British Standard combs, as, during actual build up, as many as sixteen are required for brood rearing; but in a backward or late district such as the hill country of the North of England I have had great success with only a brood chamber and a shallow super. I attach great importance to permanent reserves of food. I think Brother Adam was quite right when he said years ago, that a colony of bees required no less than forty pounds of stores at all times of the year and particularly during times of brood rearing. It is, of course, in the good seasons that the performance of the yellow-banded bee becomes apparent. Having at its disposal a larger foraging force it soon leaves the dark bees far behind. In an average season it holds its own.

Many beekeepers favour Italian bees. There are others, like my old friend the late Frank Cessford, who prefer the brown bee, or the bee which closely resembles the old English bee, although it is doubtful if many of the original lines survived the ravages of the Isle of Wight disease (acarine infestation 1918-19). Mr. Cessford found these bees most suitable for our Northern conditions. Although not prolific they had many qualities not found in the bees of the present day. Their hardiness was remarkable. With the importation of foreign bees on a large scale, the brown bee was soon replaced by cross-bred bees of various colouring and qualities, with the result that many colonies exist today which are not profitable to the beekeeper, and these should be sorted out and re-queened with tested queens of a reliable strain. In the summers we experience, Italians continue brood rearing whatever the weather, and can become a liability. At all times they require liberal feeding with sugar syrup.

This was certainly the case with the progeny of queens I have imported directly from abroad. They turn every ounce of honey into brood. But in a carefully developed strain, such as the bees used by Brother Adam and myself— or any other carefully developed strain for that matter—such tendencies are virtually non-existent. But here I would emphasise that into these respective strains certain characteristics of importance have been incorporated—by cross breeding. Colour has only been maintained at the expense of much time and patience. I have managed to incorporate into my strain the comb honey qualities of the brown bee, particularly the ability to fill and finish sections with a flawless white capping, an item so highly prized by the old beekeepers at Riding Mill. Stamina, too, has been developed to a fine degree for heather work, along with the ability to winter well on stores made up almost entirely of heather honey. This is an important item in the North-East, as we experience weeks of long, unbroken frosts. Finally, we have paid attention to the disinclination to swarm. This is an item of economic value in commercial bee-keeping today.

Frank Cessford made the following pertinent comments just before he died:

"While I do not condemn the Italian bee, which has many things in its favour, I do feel that we should pay more attention to improving the darker strains— particularly those related to the native bee of these Islands. This can be attained by introducing a certain amount of Italian blood into the apiary—to improve the laying powers of the queen and to provide a buffer against addled brood. Italian crosses of two or three generations would be most suitable for this purpose, as a gradual change in the strain is most desirable, otherwise we may soon lose the characteristics which we are trying to retain when working for comb honey."

Mention has been made here of Messrs. Robson and Cessford of Riding Mill, who maintained an extensive apiary of over one hundred colonies for the best part of fifty years. These gentlemen were fortunate to possess a strain of bee which required no more than nine British Standard combs for full development. As the use of a separate mating apiary was impracticable real progress was made by selection of the queens alone, and, as an additional safeguard, they took pains to distribute virgin queens from their own breeding stock throughout the district.

Bob Robson

In a normal season (1927 and 1954 excepted) the nine comb colonies of these bees could be expected to fill 16 shallow combs (narrow spacing)—first with flower honey, which was extracted, and then with heather, sealing all the combs. The aim was to get every comb completely sealed. Two small 8 frame shallow supers were used, and this allowed for ample packing at the fell. Or, alternatively, they supplied some fifty finished sections, one rack of flower, and two of heather honey—eighteen sections to a rack. There were rarely more than six full combs of brood in these colonies at any one time, and they came off the fell with the brood combs well provisioned for the winter.

Keenly interested in bee-breeding, I am satisfied that to obtain consistent results in honey production over the years, it is necessary to maintain two distinct lines or strains and bring these together for cross-breeding work. I have then, in one apiary, a yellow-banded line which originated from Brother Adam of Buckfast Abbey and, in another, the brown line developed by Robson and Cessford. These are brought together—invariably yellow virgins and brown drones—in a separate mating apiary and the resulting first cross goes on to head the honey producing colonies in the remaining apiaries. Situated in a heather district hives are designed for rapid moving and twenty British Standard combs are generally in use. This large comb area provides a permanent reserve of food and I encourage ample natural stores. I have the fifty-sixty pound mark in mind.

The home apiary at Shilford is in a very sheltered spot and gets all the sun there is during the winter. The other important point is being situated on light haugh land, it is dry. I am fortunate in this respect as maximum amount of sunshine throughout the winter is of great importance, as generally in North East England we are faced with long unbroken frosts. Ling honey is invariably a major item in winter stores, and some thought must be given to its peculiarities. One cannot successfully bring bees through in places which are in the shade. Losses, due to neglecting this can, I have found, be very high. Dysentery sets in towards the end of January, and colonies get into a bad way in no time. I have been obliged to reshuffle apiaries several times before being entirely satisfied. In short, good wintering here demands as much sun as one can get along with plenty of light breezes.

In the Middle Ages an old Hermit is supposed to have lived and made bee skeps in the valley of the River Coquet, and for centuries the charming custom of "Telling the Bees" at times of births, marriages and deaths has persisted in the villages. This was still an accepted thing at Riding Mill when I began my own beekeeping apprenticeship, and I am not ashamed to say that I do this even today. For, after all, who are we to withhold things from the bees?

"Trembling, I listened: the summer sun
Had the chill of snow:
For I knew she was telling the bees of one
Gone on the journey we all must go! "
(WHITTIER — "Telling the Bees")

---

(1) This is a personal experience. The author's observations for the country as a whole indicate that the dark bee gives a better all-round performance at the heather.

CHAPTER FOUR
YEARS GOOD AND BAD

" So when you see a honey bee
sit resting in the sun
you'll know that she's remembering
and smiling at the fun".
(HELEN LAURENCE. Bee Craft.)

EARLY in 1950 I was in the Riding Mill apiary of Frank Cessford for the last time. Shortly afterwards this noted apiary was disbanded, and a link of almost half a century with active beekeeping was broken with the retirement of its owner. For many years Robson and Cessford dominated Northumberland beekeeping, and with the exception of the Bamboroughs of Alnwick, were our most extensive honey producers. It is to the senior partner, Bob Robson, that I owe much of my own interest in the craft. So great was this man's love of bees and so skilled his management, that I count it an honour that he continued as my mentor in the field of practical beekeeping until his untimely death in 1952. Frank was philosophical in outlook, and it was his delight to pass on to the uninitiated a wealth of bee lore.

The years which preceded the twentieth century are regarded indifferently by beekeepers, and it was not until 1911 that the sun shone out at last, in a way which would always be remembered by those who lived through this halcyon year. 1911 is imprinted in the minds of farmers as the season of heavy morning mists which slowed down hay-making until afternoon. But for bee-keepers the honey flowed. At the fell the nectar flow was so protracted bees came out from their hives and built combs underneath the floorboards [1]. Isle of Wight disease (later known as acarine infestation) made its first appearance and soon became rampant throughout the British Isles.

The years 1912, 1913, 1914, and 1917 were disastrous to British bee-keeping, but 1915 was a remarkable season, and 1918 and 1921 are remembered as the seasons with the heaviest clover flows known by Northern England, yet both these years were surpassed by 1955 and 1959. Lean years were to persist from 1921 to 1930, with 1927 receiving mention as one of the wettest. The sun shone once more almost from 1930 to 1935— a unique run of seasons which had an embarrassing effect on the marketing of honey. The position was alleviated by a run of poor seasons, and the marketing problem righted by the outbreak of World War II.

About this time I established my own apiary. As I remember them, 1943 and

1945 were both good seasons; a difficult 1946 was followed by the great winter and the magnificent season of 1947 with its unusually early heather flow. Up to that time it was the heaviest known—and all gathered by August 14th. The year 1948 was disappointing in the extreme. Then we came to 1949—again one of the great seasons in British beekeeping. Bee colonies never looked behind them and my own stocks were on the fell by July 20th. 1949 still has the distinction of having the heaviest flow of heather honey. From one apiary of twenty colonies, situated at Harwood Shield [(2)], Hexhamshire, I obtained a crop of 1 ton, 1 cwt. of heather honey.

1949 - a great honey year. Colin Weightman in his apiary at Blanchland.

With the exception of 1951, the years which followed 1949 were in keeping with the 1920's. 1954 was probably the most disastrous known. At last 1955 came to us; after a barren spring, suddenly, in June, the wind changed to the East, and from 28th of that month honey poured into the hives. Supers were piled on all colonies, but it became increasingly difficult to cope with the flow, which finally terminated on July 29th. On the following day I removed 208 Ib., 204 Ib., and 186 Ib., of clover honey from three test colonies in the home apiary. The many visitors to the Northumberland Bee-keepers' Field Day that year, saw these stocks at work.

1956 was a peculiar season all round. When, as last, the long dry spring came to an end there was certainly rejoicing over the whole of North Northumberland and the Eastern parts of Scotland, where things were really desperate. Day after day, encouraged by an East wind, great clouds of dust left the fields and spiralled high into the heavens making working conditions most unpleasant. It was quite the driest time I can remember. The heather season too was one of the most difficult. "We have to go back to 1912 to find a year to compare with this one." This is what one of our leading hill farmers told me when I came inbye from the fell to shelter after feeding bees. I had never encountered a year quite like 1956, so rather philosophically my friend said that it reminded him of the period

Colin Weightman in 1955.

1910/1912; 1954 could well be compared with the disastrous 1910; 1955 with the glorious 1911, and 1956 with the fateful 1912. From mid-July there was hardly a day when rain had not fallen on the fells, more often than not incessantly for days. I got the bees out under conditions which were far from ideal. Clouds lay low about the hilltops, dank mist covered everything, in fact the only sound came from the brown waters of the burns and becks in spate tumbling down the ghylls.

For two weeks I was feeding heavily, and it was quite a business journeying to and from the fells with syrup. I took the opportunity to look in at a few communal stances owned by our Association members and at every one some colonies had gone down. Fortunately I had a supply of spare feeders with me and where bees were just starting to roll out of the hives I put these on, and I believe this kept the colonies going until the owners could get out to them. On the few occasions when the skies did clear, gales of great force made the Northern moors most uninviting, and there were hardly any flying periods at all.

The Editorial of "The Field" for this period refers to a fell where I had forty colonies (twenty to a stance) and says "there was such a hurricane on August 13th that it was almost impossible to stand." But I was disturbed most by the lack of bloom. This part of the North East, usually regarded as one of the driest places in the country, with an annual rainfall of little more than twenty-seven inches, broke all existing records for the August of that year with a reading of 9.33 inches. Indeed, we have to go back to 1855 to find anything to compare with it. In that year 7.29 inches was recorded here in August.'

My farmer friends aptly referred to it as "heart-break harvest". Many of their problems are shared by us, and it is hard to disguise the dismal picture when visiting the fell. My fleeting visits to the heather stances were directed towards removing empty section racks—and you know what that implies when bees are loathe to pass the "clearer boards" but drenching rain means no honey. The veteran, Mr. Tweddle, of Riding Mill, told me that in his fifty-four years working of the local fells this was the first occasion "when they failed to put something in the bottoms" (referring to storing in the brood combs). "They always got something—even if it was into September" [3].

Then dramatically in mid-September conditions changed. The calamity foretold did not materialise and I was able to report some intake from the ling. Although all available space in "the bottoms" was filled, there was a notable absence of new white wax, and consequently, any attempt on the part of the bees to seal. But the point for which we were profoundly thankful was that they had their winter keep. After all those weeks of deluge I caught for one fleeting moment the splendid scenes from former years, but hastily realised my plans were for next year.

As in 1954, those who sent their bees out with a permanent reserve of food, and latish (mid-August), got something. But for any real assessment of the situation we have to distinguish between what was already there and actual intake from the ling. In comparison the colonies which had no reserves of food (the section producers were in this group) were in rather a poor way. My late re-queening was slowed down and I had more nuclei than I am used to standing in apiaries. Despite the hazards of July and August, they responded well to feeding and were in nice fettle for the winter.

In 1957 the wild cherry made one of its best shows for years and plant growth was even earlier than in 1949. Travelling through the English countryside on those lovely spring days one was struck by the intensity of everything, the blueness of the sky and the vast panorama of verdant green. Browning's lines, "Oh, to be in England now that April's there," sprang to mind.

We were kept extremely busy on the farm, seeding being completed under the best conditions for years. And, of course, we had assembled frames and fitted and wired foundation. I like plenty on hand, and at the rate at which colonies developed many of these were in use within a week or two.

Mr G Tweddle

Mr. G. E. Tweddle (Riding Mill), summed up the state of affairs in his own inimitable way when he said casually one day—"They're blowing it in."

The previous autumn I went so far as to say that after all the sunshine we might look forward to some "tree honey" and in my part of the world this materialised. Day after day, after a dull start, the face of the County became steaming hot by midday. As the sky cleared and became a brilliant blue, the massive shoulders of the Cheviots stood out to the North, their outlines clearly defined. Then there followed a sizzling heat; the countryside slumbered, but amongst our beekeepers there was great peace of mind.

I am indebted to my friend Professor H. C. Pawson (Emeritus Professor of Agriculture, Durham University) for the following information: taken from the Cockle Park Report where weather records, etc., have been kept since 1898. These were years when surplus honey was obtained in quantity in Northern England.

EXTRACT FROM COCKLE PARK REPORT:
SUNSHINE (HOURS)

1901 is reported as having had 1,746 hours of bright sunshine: but 29.76 inches rainfall.

YEAR	1911	1921	1943	1945	1947	1949
JAN.	52	35	52.60	62.10	37.20	68.20
FEB.	85	36	111.10	93.90	45.90	115.00
MAR.	112	100	129.70	113.90	90.90	110.20
APR.	104	198	161.20	184.60	161.60	145.80
MAY	227	241	209.20	167.90	164.40	211.70
JUNE	216	213	199.80	193.60	122.00	246.60
JULY	245	166	202.70	188.20	177.40	178.80
AUG.	189	120	111.50	145.10	266.80	141.60
SEPT.	198	164	148.90	162.30	266.80	120.90
OCT.	68	125	80.70	108.10	80.40	89.10
NOV.	69	69	75.00	50.00	110.00	54.10
DEC.	43	51	45.70	47.00	52.50	44.60
TOTAL	16,608	1,518	1,528.10	1,506.70	1.414.90	1,526.60

Average for fifty-six years at Cockle Park: 1385.17 hours and 28.678 inches rainfall.

(1). It has been suggested that a wet May is essential for the success of the Ling later in the year. The author finds that hours of sunshine in Aug./Sept. matter most.

(2). Bees in this apiary occasionally work the Rushes (Juncus communist) before turning their attention to the Ling.

(3). Occasionally in very wet years the heather beetle (Lochmaea saturalis) devastates large expanses of Ling. Fells with a reputation for being wet are usually the worst affected, while the dry fells escape. A well known Lochmaea year was 1946 when Bollihope moors (Co. Durham) were badly attacked.

CHAPTER FIVE
HIGH SUMMER

"... Stands the church clock at ten to three ?
And is there honey still for tea ? "
RUPERT BROOKE

SEASONAL work at Shilford takes this form. During late May we carry out a general examination in all apiaries, and colonies found below standard strength are given combs of brood and bees from sources that can spare them. The emphasis is upon equalising strength throughout the apiary. At this time any colony showing signs of exceptional development is provided with a super. In June honey storage room is provided for all colonies, and any not now up to standard are re-queened by uniting nucleus colonies containing tested queens, which I overwinter in the sheltered queen mating apiary. The month of July sees further storage room provided where required, and a general re-queening of all colonies with young queens in preparation for the heather harvest [1].

At Ted Humphreys with Mr S. J. Cox

I bring my feeders into use only when some of the single brood chamber colonies appear to be getting really short of stores. Stimulative feeding should be avoided here. This is similar in many respects to the pattern used by Ted Humphreys, my commercial friend in Lancashire. One of the problems we have to face is to provide a feeder of reasonable capacity, which can be warmly packed, and the syrup must be readily accessible to the bees so that it can be taken in an hour or two if necessary. The larger "tray-types" which hold a gallon or more of syrup, are unsuitable in this part of the world—and, in fact, the only time they can be used with any measure of success is at "fell-time." When these leapt into popularity just after the war, I had a number in use for spring work, but during a cold May and June I lost some really forward colonies because the large colonies, when short of stores, would not take the syrup down. Mr. Hodgson Gray, in the North of the county, was faced with the same problem and at a Highland Show we thrashed the problem out. As a result he placed an order with Steele and Brodie for an all-wood feeder, which would fit nicely into a super and would allow some packing.

Queens' flights have always interested me. In March, 1953, I watched a marked breeder queen on such a flight in the farm apiary. I first found her on the grass. Soon afterwards she flew and alighted on a hive-top, and here she went through a whole rigmarole of movements which could be likened to the "face washing" activities of the worker, finally evacuating a few small drops of an almost clear substance. After ten minutes sunning she again took wing and I watched her enter a hive in the group where I was standing. Subsequent colony behaviour was quite normal.

In the giving of a second brood chamber, or super, a good guide for Northern parts is to wait until the hive is full of bees and the cell mouths are being extended with new white wax. This may be too late for some districts: but much harm can be done in these parts by providing backward colonies with too much room early in the season. It is usually advisable, when bringing a second brood chamber into use, to raise two combs of brood from below and put them together in the centre of the new chamber and their places being taken in the lower chamber by frames of foundation—not together but interspaced [2].

With the merry month of May the sycamores, which abound in the parkland near my home, come into flower, and the attention of the bees rapidly turns towards them. Although in high lying districts it will be some time yet before they provide a flow, it is surprising the number of our hillmen who are migratory in their beekeeping and come down to the valleys. When the sycamore has started to secrete nectar I like to go out into the home apiary at night and by stooping down amongst the hives, listen to the roaring and attempt to catch the characteristic smell. Hawthorn, too, will be out in sheltered places, but more often than not the bees leave it severely alone. During heavy hawthorn flows bees will be seen hanging out of the hives. The Cumberland people sometimes do well from it. Areas of bluebell now show, and the banksides colour overnight. Every year I find a few colonies storing a particularly light coloured honey. Mr. Cunningham of the East of Scotland Agricultural College once told me that this could be from the bluebells (3).

Messrs Cunningham and Weightman

The colonies on single brood chambers should have come on nicely, and I take advantage of their forwardness to get some brood combs drawn out, as I find there is no better way except perhaps in nuclei of securing good even combs, than by using this flow.

Eight British Standard frames fitted with foundation (wired by myself in more leisurely moments) are given to each colony in a second chamber and soon they should be drawing the foundation out well. But it is often necessary to move the outside frames further in to get all the combs completed.

I have now to give some thought to queen rearing, but this is largely influenced by the current prosperity of the bees.

Early in March I distribute ten (drawn out) shallow drone combs—with some stores—in the food chambers of those two colonies earmarked for drone production: the surplus combs being removed. Towards the end of April I move these colonies out into the shire for their summer vacation in the mating apiary. Incidentally, the queens here winter splendidly. I have mentioned that I do not believe in examining colonies in the apiaries until the latter part of May. Then new wax is in evidence and the cell mouths are again being lengthened.

Swarming is usually in evidence in the North from Whit week-end, and some very good swarms come off too. These early swarms are usually set up on a new stand, and after due examination, one, or sometimes two of the most promising queen cells are left in the parent colony [4]. The bee-keeper following this practice is invariably able to send two good colonies to the fell. But one thing has disturbed me: there appears to be an increasing tendency to use drawn comb to hive them on. When I worked bees this way I liked to use as much foundation as possible as this reduced the urge for building queen cells. I mention this as many of the swarms I was shown that year (hived on all drawn combs) went straight ahead with queen cells again, and they in turn swarmed once more.

Talking about swarms reminds me that I meet men who find that much of their bee-keeping revolves round this phenomenon. One of the things which has come to light concerns the behaviour of the swarming colony when confronted with a cold snap. How does the colony deal effectively with the situation, and destroy a batch of queen cells which are sealed ? Sometimes one discovers a colony with sealed queen cells (where the bees have not come off) and find upon examination, that all the queen cells which appear normal contain dead larvae. I hold that these have in some way been chilled, but my "veteran" friends dispute this, and inform me that they have had many instances of destroyed queen larvae in colonies boiling over with bees where there was no chance of chilling. I never get this in my planned "bee-breeding" and wonder what is the general experience of beekeepers. Do the bees themselves destroy the queen larvae in these "swarming colonies"—perhaps by stinging when bad weather intervenes?

My Gamekeeper called me out once to take a top swarm which had settled well down in a thorn hedge. We spread a few sacks out below the swarm, turned the skep mouth upwards, and with the aid of a heavy post dislodged most of the bees, and the best part of them went into the skep. We turned the skep over, settled it on twigs and it wasn't long before the stragglers joined them.

I was surprised to find when we came to open up the parent colony, that all the queen cells were unsealed, and there was much unsealed brood with too few bees to cover it. This will give some indication of the heat we experience on occasion in the North. In this particular case we hived the swarm on a new stand, on frames fitted with foundation and left the parent colony intact, as my friend intended to watch the position carefully during the next few days. Cell cups were numerous, in fact I counted twenty-five, but on one occasion we removed over forty queen cells from ten British Standard combs!! It is important to hive the bees on frames fitted with foundation only. Hive on drawn-out comb and you may have a maiden swarm, out within a fortnight—egg-laying having taken place amongst the new combs queen cells are started without delay.

I liked Taranov's (Russian worker and Editor of a Bee Journal) suggestion that when swarming bees filled their honey sacs before flight they stayed with their queen; but those that didn't returned home. I think my Gamekeeper's bees belonged to this category, as many bees appeared to leave the swarm and return to the old site, and in consequence the unsealed brood. about which I was concerned, took no harm. Brother Adam in a lecture (Seale-Hayne Agricultural College, Devon, 1929) discussed a method of swarm control which could usefully form a basis for any intensive system of management even today. For a full description see British Bee Journal, August-September 1929. But for all who find time and years against them I recommend the method put forward by my old friend Mr. L. E. Snelgrove in the 10th Edition of his "Swarming—Its Control and Prevention," This is referred to as his method IV.

Occasionally we get fine weather in June as the following diary extract shows. "Two weeks of glorious sunshine have convinced me that the old-timers were right in their accounts of 'flaming' June. Optimism prevails over the whole Northern scene. Bees got straight away on the white clover and I have experienced one of my earliest major flows. This was so prodigious that I brought all available supers into use. But there are reports of colonies being in poor shape with build up now instead of earlier: a hangover from last year. However, there is still a good chance that these will be just right for the fell. Incidentally, the heather is likely to be early as corn is shot in many places, and the heather and our corn harvest go hand in hand.

"When the nectar's pouring in.
All day long a merry din.
Supers added more, and more;
Then at night you'll hear them roar."
(A.R.C. — Bee Craft).

(1) The opportunity is now taken to clip the wings of all mated queens.

(2) With adequate breeding room, and a natural barrier of stores, queen excluders are not in general use.

(3) Bluebells (Hyacinthoides non-scripta, not to be confused with Harebells, Capanula rotundiflora).

(4) The deciding factor is the strength of the parent colony. If strong in bees, only one queen cell is left. But should the population be depleted, it is safer to leave two queen cells. In the case of small apiaries the problem can be overcome by division of the parent colony into two, or more Nucleus colonies (with one queen cell in each). Following mating, the best queen can be kept, the bees re-united, and the remaining combs utilised as required.

(5) Those interested in the conditions which lead up to swarming should become familiar with DR. COLIN BUTLER'S work on Queen substance. Reports indicate that it can be manufactured synthetically and the possibility of it being supplied to bee colonies during the swarming season cannot now be overlooked.

CHAPTER SIX
HEATHER-GOING

"Wild blossom of the moorland, ye are very dear to me;
Ye lure my dreaming memory as clover does the bee:
Ye bring back all my childhood loved, when freedom, joy, and health
Had never thought of wearing chains to fetter fame and wealth ".

ELIZA COOK.

PROBABLY the most rewarding experience in beekeeping is to secure a good crop of heather honey in sections. The flawless white cappings of the cells, the rich amber colouring, the texture, together with the pleasing aroma, all are contributory factors in making it a most prized and unique product of the hive. It carries the tang of moorland air—a honey which can command its own price in the face of all competition.

During the 1939/1945 War, Mr. Jasper Stephenson and my father looked after the War Agricultural Executive Committee's ploughing-out scheme in the area: and when, in my very early days of bee-keeping I became "heather minded", Mr. Stephenson went to great pains to ensure that I got my colonies to the right spot. Little did I know that the Acton, Edmundbyers and Ladle Wells areas embraced some of the best fells for honey production in Britain; and this has been brought home to me time and again in succeeding years. When the heather crop has been a failure in other parts of the country, these fells have never let us down. Probably the fact that they are so deeply incised, have such contrasting moorland scenery, and the heather itself is handled properly accounts for it.

After allowing colonies the run of two brood chambers, or a brood chamber and shallow super for most of the summer, we like to send colonies to the heather on a single brood chamber only re-arranged with, say, two combs of stores to the outsides of the brood, which, by the way, is mostly sealed. Our small section racks holding only 18 to 21 sections respectively, are then placed in position and packed around the outsides to ecourage complete sealing of all sections. We aim to use as many drawn-out sections as possible for heather work. If you have a few unfinished sections (with ling honey) left from the year before drop them in, as I am sure this helps to get the bees out on to the heather. The colonies in the queen mating apiary [1] provide me with the best indication of the trend of things; and the final decision to move the honey producing colonies is taken when the first ling honey is being stored (up here) in the combs of the nuclei.

I aim to go up into our hill country in July just to see what is happening, and more often than not find the ling to be rather on the late side. A controversial point amongst heather men, it is claimed that the great heather seasons are usually early ones—when the ling is in flower in July. Harwood Shield in the Hexhamshire area described by the late Bob Robson of Riding Mill as being probably the earliest fell and best in the North-East, is usually green, and is always a fortnight earlier than any of the others. But then if you are a real connoisseur of heather honeys it is claimed that the late fells are the best and that the honey has more body.

I know the late Alf Dodd, the well-known comb honey man from Hexham, thought along these lines. Climbing steeply from the old town of Alnwick the moors here again are early ones; the Simonside hills make a splendid vista in the evening sun as you come back into Tyneside via Rothbury and Weldon Bridge.

Despite the practical trials, "heather-going" can be an inspiring experience. On those glorious August and September days when the countryside lies so benignly in the sun with the fells aglow with colour it is possible to achieve a unique peace of mind and sense of satisfaction.

The usual excitement prevails as hives have to be made secure. Many gadgets are available for bringing the parts together, but apart from crate staples in floorboards and brood chambers, I prefer to cord the hives, using a short peg of wood as a tourniquet to draw in the slack (here is an outlet for the accumulation of baler twine which becomes an embarrassment on many farms). The East of Scotland Agricultural College make use of a banding machine, and Mr. and Mrs. Leech of Swaffham (Norfolk) when they once visited us here gave me their device for re-using metal bands. Entrances are closed entirely with neatly folded bags, but top ventilation is provided. Mr. W. W. Smith in Peeblesshire uses lengths of felt.

Stances must be chosen with care, and it is only after a number of years of experience that a true evaluation can be made of a fell. My own knowledge of local fells is in fact only scant. But the one thing which has been clearly demonstrated to me over the years is the fact that you require no more than twenty colonies to a stance, although stances need be only half a mile apart. Taking all years into consideration the twenty colony groups covering an entire fell give a much better account of themselves than the communal groups—made up of many colonies— so often seen. Serious heather-honey production presents a complex problem. On our wind-swept [2] uplands many factors, often closely interwoven, come into play, the most pronounced being climate and topography (rainfall can be heavy in the higher valleys and only ten miles distant be virtually non-existent). Our fells being deeply incised, protection of the valleys must be sought. One may well surmise that the ling on one fell is a prodigious yielder, but with a change of wind the flow stops suddenly at a stance, but visiting the next stance, where the bees have up to this time been inactive, a heavy flow will be in progress.

The first snow of the winter can be expected in October and the Lakeland mountains and highest outcrops of the Pennines gently capped with white, make a scene, which, if nothing else, is vividly picturesque. Then a bitterly cold wind blows straight down from the North, and here in North-east England the leaves change overnight and Autumn in all its glory is with us. My last visit to the heather stances is made on a bright cool day when the bees are still flying well, and the opportunity is taken to "cord" the hives ready for lifting. Here, as in the valleys, the sharp night frosts have changed the colour of the bracken and at first sight the whole expanse of moorland appears red.

At home the bees continue to take down syrup well. I feed and feed colonies just back from the fell until the first killing frost arrives. I agreed wholeheartedly with the late Jim Cox over the merits of having large slabs of candy for all the smaller lots of bees (colonies without a permanent reserve of food). This is certainly much safer than feeding small amounts of syrup throughout the winter— no good has ever come of this in my hands, as it promotes undue activity at the wrong time of the year.

Mice have by this time already begun to leave the fields and I pay due attention to the hive entrances. I have already got down to placing entrance guards in position; these are pieces of small mesh wire netting fixed with drawing pins.

I had this device indirectly from Mr. Manley some years ago, and like it, as it is an improvement on the "Queen excluder on floorboards" business I used for years. But the point is, I have reduced the numbers of queenless colonies in Spring, and I am now quite sure that the excluders interrupted flight of queens in early Spring and caused undue excitement.

Sir John Craster, one of Northumberland's leading naturalists, once remarked that we could, over the years, look forward to three well defined spells of weather These could be expected to turn up faithfully year after year,—the cold snap of mid-February,—the "blackthorn winter", —and St. Luke's little summer. But more often than not the weather remains open, the sky is of the softest blue, and after heavy morning dew the fields and hedgerows appear carpeted with gossamer. The bees now take syrup fast and on every warm day one can expect to see a deal of pollen going in. I am pleased when colonies respond to attention and the sight (when bedding down) of some nice compact clusters tends to relieve many of the qualms I have at heather time.

One of the privileges enjoyed by those living in the country is the opportunity to observe nature at close quarters throughout the year. Local personalities list the following as indications of a hard winter:
- when the Ash leaves have hung until the last lap before dropping;

- the brown trout have moved up into the becks and runnels fully a fortnight before their time to spawn:

- and again, the squirrels (the red kind) have behaved prodigiously in harvesting. Their efforts among the beech mast are surely among the most arresting sights of all.

As for the bees, it often happens that hard weather follows in the wake of a difficult year—just to try you. But I have personally noted that time and again persistent November fog invariably presages a good honey year.

Those of us in heather districts know only too well the boost colonies derive from ling honey. Colonies so provided give the best account of themselves the following year, and for some time now I have set aside a super of ling honey for all the honey producing colonies, and I am convinced that this is one of the best things I have done.

When our ling honey is obtained in such quantity and at a time when admixture from other sources is small, then conditions are against crystallisation. I instance 1946 and 1954 when the flow here was in September and entirely from the ling, and it was virtually impossible to induce crystallisation by any means at our disposal. Mr. J. Pryce Jones obtained some of this honey for his work. I see Brother Adam mentioned a similar experience when lecturing at Newcastle in 1934.

In a good heather season we aim to remove the completed section racks at the fell to help those who go to lift the hives, and here great care must be taken not to use too much smoke or things like the carbolic cloth, as much comb honey is spoilt through tainting. But more often than not the clearer boards are put on at home and in the out-apiaries. All the sealed racks after a light scraping go into our store which is a warm, dark room, and storied to about 12 high with occasional clearer boards between them, and on top of each group. They are taken out as needed and the sections prepared for marketing. We were still selling combs of 1949 vintage in 1954. But there must be adequate protection from mice and the ravages of wax moth. We have always relied upon P.D.B. crystals here and they have served their purpose [3].

In recent years, there has been a tendency, particularly in Scotland, to depart from sections altogether and produce first class comb honey in shallow frames, ekes, and cogs. This honey is chunked to a required size [4], allowed to drain, and wrapped in cellophane. The addition of a piece of tartan or spray of heather will tempt the wariest of shoppers and the high standard of presentation captures the exclusive tourist trade.

Other producers in heather areas favour the standardisation of equipment and in many apiaries brood chambers only have replaced shallow supers and section racks, the one lot of combs simplifying the task of the operator. The honey is usually scraped from the midrib or the comb cut entirely from the frame for pressing. Others favour the use of instruments such as the "perforextractor" or honey loosener which has come in from Norway. The heather press marketed by Mr. Abbott of Mountain Grey Apiaries, Yorks., gives an impeccable performance, and where there is a substantial crop to handle, two or more can be used with advantage. Everyone should aim to see the really celebrated press installed by Brother Adam at Buckfast Abbey.

When I visited the West of Scotland Agricultural College, Auchincruive, in the Autumn of 1959, Mr. Savage showed me the "Yarrow" hive developed for section honey production in the area. Standard brood chambers are again utilised and the sections themselves are suspended in a metal framework faintly resembling a basket.

Honey from the Bell heather is rarely obtained in quantity on the Northumbrian moors although many Scottish beekeepers secure a crop from this source. During a hot July the combs in my queen mating apiary are often well filled with it and sealed. But you must acquire a taste for it.

During our beekeeping apprenticeship we subscribe in thought to many ambitious schemes designed to solve the problem of the marketing of our crop. But sooner or later we are brought face to face with harsh reality and realise how nice it would be if we could supply all our customers—large or small—with their usual quota of honey. After a succession of cold, sunless summers, suddenly there is very little honey about and unless one is in the fortunate position to hold an extensive crop in store for such emergencies there is very little one can do about it. You then realise how difficult it is over the years to keep even a modest supply of honey flowing into shops throughout the country.

So far as flower honey is concerned the marking scheme operated by the Bee and Honey Associations in the 1950's, in conjunction with BSI, served a very useful purpose and found favour amongst a number of small producers in this country. It replaced the National mark of pre-war days. Now that BSI operate the scheme alone, many hope that it will eventually be extended to embrace the heather honey producer too. Those really interested in a standard for heather honey should see the excellent papers by Mr. Alex. Deans which have appeared in the Scottish Beekeeper.

1957 was an indifferent heather season—with one very hot week at the beginning of August, when bees got all there was to get that year. Wind and rain persisted for the remainder of the time—the peak blooming period—so, naturally, there was little honey about in these parts and almost none in Scotland. But Mr. A. F. Wheeler told me that the bees had done well in the New Forest areas of Hampshire, and a similar report was received from Mr. Leslie Hender from the Quantocks (Somerset).

The position was reversed in 1958 when the North East was one of the few places in the British Isles to secure a crop of heather honey. An "Indian Summer" (August 29th/September 5th) coincided with the peak blooming of the ling and the wind settled nicely in the East. On the fells of the Durham- Northumberland border (1,500/2,000 ft.) a tremendous nectar flow was in progress. Bees were hanging out of their hives to such an extent that the faces of the brood chambers were covered to the first super, with a further mass of bees spread out upon the boards I placed before the hives. Section racks were almost completely filled and sealed. I was obliged to provide some extra room, but I am always chary about this when it gets late on in the year. This was the heaviest (nectar) flow since 1949. The heather flow of 1955 was protracted, rather than heavy and of short duration.

Some painstaking work should be done on wintering bees on heather honey and on the peculiar qualities of the honey from the ling. My friend, Jack Cox, who is responsible for horticulture in Co. Durham, with headquarters at Houghall, has made a special study of the heaths and all who are attracted to this fascinating subject should aim to get in touch with him.

In 1959, we experienced a season which was probably unique in the annals of British beekeeping. To find its nearest counterpart we must go back to 1911. The sages said it had to come after all the previous winter's fog and the long run of cold sunless summers, but we hardly anticipated what we got. Bee colonies never looked behind them, and in fact there were only two days from when foraging began in earnest in mid-April, that they were confined entirely to their hives. Every other day provided some opportunity for them to go abroad if only for a short time. Once the active season got under way queens were laying in no time, combs were drawn out overnight, and whole section racks were filled and completed in a week from starters of foundation. The ling heather was in flower on the fell by mid-July and we had the bees out soon after the 20th. The flow started almost at once and went on and on; for those who were able to remove section racks and supers as they were completed (about once a week) it was possible to harvest a remarkable crop of heather honey. We finally brought the colonies off the fell on the 16th September and after removal of the crop they were well found in the bottoms. Those who lived and worked through those August and September days, when the mist lay low in the valley bottoms until noon, and the temperature then soared into the 80's, will cherish a picture of the English countryside in all its pristine glory. A harvest [5] both on farm and apiary never to be forgotten, and the sight of the inhabitants of these Islands browner than they have ever been before.

(1) 1947, 1949, 1955 and 1959 (all hot summers) we obtained honey from the Scots Pine in this moorland apiary. The Scots Pine flourishes in many plantations which provide wind-breaks on the Northumbrian fells. Scottish bee-keeping friends regard it as a poor food for wintering. The author's experience is the opposite, but is limited to one apiary only.

(2) At the turn of the century MAJOR F. SITWELL (Ainwick), Secretary of the old Northumberland Bee-keepers' Association, coined the following saying: "When the wind is in the west, then the flow is at its best."

(3) MR. ANDREW STEWART has perfected an instrument which makes a first class job of chunking heather comb. MR. C. BRUCE, Pencaitland, describes this in Scottish Beekeeper, 1960.

(4) PDB crystals are no longer recommended for wax moth protection as they can leave residues in the hive and are said to have cancer-causing properties for honey consumers. (Editor)

(5) On this particular occasion bees in the author's home apiary at Shilford (recently brought back from the heather) were observed working the wheat (Triticum) stubble in large numbers for about a week.

CHAPTER SEVEN
A WINTER'S TALE

"There's a whisper down the field where
the year has shot her yield,
And the ricks stand grey to the sun
Singing: ' Over then, come over, for the
bee has quit the clover.
And your English summer's done'."

RUDYARD KIPLING - "The Long Trail"

THE fall of the leaf. Mellow autumn days with mists. The sight of lea and stubble turned in readiness for winter's storms has a magical attraction of its own. It is at such times that the countryman can sit close into the fire hob at night to converse and solace his soul. Traditionally the bee-keeper drinks mead. In 1953 Mr. Brian Dennis, Harrow, Middlesex, gave an important paper (as a lecture) to The Central Association of Beekeepers. This was "A Background to Mead Making." Mr. Dennis cut through the veil of doubt and presented in concise form the things one should know about the subject. As a young man, Brian Dennis had close associations with Newcastle's Armstrong College (now King's) and he has maintained an active interest in Northern beekeeping ever since.

Taking full advantage of the mild days when they come I make my customary New Year inspection of the apiaries [1]. But I like to wait until the sun is shining strongly. Then the bees from almost all colonies should be flying well. No matter what the future holds in store, they stand a reasonably good chance of coming through. I recall that in one apiary in the shire (at 1,000 ft.) two colonies were in poor shape, seemingly queenless, but still with ample stores. We waited until the sun went in brushed what few bees there were from the combs with grass, then closed the hives. This, to my mind, is preferable to letting them go down and have bees decomposing in the combs, with robbing on top of this.

Other mild mid-winter mornings are spent in the home apiary removing tops to encourage the "warming up" of hives, but you must always be about when you do this. The weather sometimes is a treat, and we are extremely fortunate when we get spells like these, providing we keep the stores position continually in mind. The activity is sometimes so great here that I just wonder if queens fly in mid-winter. I have had them out in March.

The nuclei, which I overwinter, each receive another block of fondant (5 Ib.). The only worry I have at home is with mice when they make their appearance in the bee-house. When the three colonies in the bee-house remain inactive I bring the hives outside. The sun is unable to penetrate here, and unless this is carried out faithfully throughout the winter colonies go down.

During my round of the apiaries, I am always struck by the qualities of the line of brown bees developed by Frank Cessford and his partner at Riding Mill. When this old Northumbrian beekeeper gave up, he asked me to hold on to his breeding stock. As in the remote parts of Northern England where the popular Italian strains fail, this bee had acquired characteristics of immense importance, and for comb honey production was unsurpassed.

I set these up in Tynedale in two apiaries entirely on their own. And they have since provided much useful material for my own breeding work viz., crossing with the yellow banded Buckfast stock. These small colonies (in fact they only occupy nine British Standard combs) stand up to our winters better than anything I know. On those occasions when the larger colonies of imported yellow bees get into a parlous state they stop brood rearing with the result that the brood combs are well provisioned, and there are some grand compact clusters settled in the midst of plenty.

In 1947 bees were confined for weeks on end, and by mid-March we found it necessary to dig the hives out and clear the ground of frozen snow for a few feet around them. These cleared patches were covered with dazed bees, but with the heat of the sun they quickly recovered and took wing. There is much to be said for this old North of England practice, especially when bees have been confined to their hives for periods of two months or more.

The question of shading hive entrances from the snow is often foremost in my mind, but the only time this presents a problem is when there is a really heavy fall of snow during March or April. Then breeding is well under way and bees do tend to fly out and become numbed, and I think we should prevent this if at all possible. I confess I have never yet managed to put this into practice in the out-apiaries. In the home-apiary I do get around to raising the alighting boards (loose boards reared against the hive).

The Westerly winds and warm air currents from the Atlantic quickly find their way through the Tyne gap to the North-East coast, and almost overnight the countryside can clear of snow.

I have rarely known colonies in better condition than in the spring of 1959. Warm and cold spells were nicely interspaced, but the striking feature was the state of floorboards. We nearly always get a heavy accumulation of small stuff from the crystallisation of stores, but this was non-existent.

I aim to go out to the mating apiary about this time of year just to see that all goes well and I like to find the bees clustering really closely, with ample stores in the outside combs. Another point I am quite satisfied with is that in all high lying districts colonies should have some form of additional protection other than that provided in the valley bottoms. I know my friend, Margaret Logan (North of Scotland Agricultural College) will bear me out here. Miss Logan takes her stand by the 12 and 15 British Standard comb "Glen" hive.

My own experiences on the Northumbrian fells revealed that colonies provided with a second outside casing, which could be loosely packed with moss, dry leaves, etc., and had plenty of top clearance over this porous packing, invariably gave a better account of themselves than the unpacked colonies [2].

Winter provides time for reflection. Past and present work can be considered and plans made for the future. The much loved farm house, fount of all activity, means more to us now as we come indoors at tea time from the frost and fog. Nostalgically we associate the wholesome smells with baking and wood logs crackling cheerfully in the grate.

Tradition has it that the former inhabitants of the Wall country brewed a famous Heather Ale. But the coveted recipe has been lost. I hope Dr. Allan Birch, who is so interested in this subject, will be able to throw some light upon the matter.

> *"Oh! the old familiar faces*
> *We've missed them for a time,*
> *Snugly housed within their dwelling*
> *From storm and snow and rime."*

HELEN LAWRENCE — *Bee Craft.*

1. Inner cover boards are in use, and a single folded bag covers the large feed hole throughout the year. When section racks are in position, quilts come into use and top and side packing is provided. With the exception of the Nuclei full width entrances are provided (with mouseguards in winter).

2. With packing, the "warming up" process is slower and many people are not attracted to it.

CHAPTER EIGHT
ROYAL JELLY AND ALL THAT

'A maiden in her glory,
'Upon her wedding day,
Must tell the bees her story.
Or else they'll fly away".

DURING the Nineteen Fifties the advertisement columns of many newspapers and periodicals brought royal jelly to the notice of the public for the first time. Colourful accounts of rejuvenation and claims that it was the panacea for practically all ills were often thrown in. But the claims for the fabulous queen bee milk were fascinating and there is now genuine interest in its possibilities in the treatment of leukaemia.

In this chapter I discuss the part played by royal jelly from altogether another angle—the rearing of queen bees for replacement purposes in my honey producing hives.

My constant aim has been to reduce manipulative work—in queen rearing—to a minimum. At a given time (usually early June) a prosperous colony in the home apiary (if this is not available a colony is brought in from an out-apiary beforehand) is made into an unbalanced unit, with a huge nurse bee population, and a temporary accumulation of larval food. Later on I want a powerful colony well provided for, entirely engaged in the care and nurture of the selected larvae destined for queens. When 30/60 queen cells are required, a colony occupying comfortably two brood chambers (National type) and twenty British Standard combs is made queenless and. broodless The entire population is confined to a single brood chamber with the comb position as follows:— loosely fitting dummy board, comb of stores (preferably honey with current season's pollen), space to take a cell-carrying frame, comb of stores, space for cell carrying frame, comb of stores and dummy board.

The queen can be held in a small nucleus colony. All other combs of brood without adhering bees are superimposed in a brood chamber above a suitable colony to be cared for. That is a colony strong in bees. Should it be proposed to treat the nucleus as a new colony, or alternatively the surplus combs, these latter can after a four hour interval, be removed to another apiary with their new covering of bees. Three hours after depriving it of its queen and brood, the cell starting colony, after preliminary distress, will have settled down to clustering notably around the entrance. It is advisable to have the "starting" colony isolated from queen-right colonies, for it is no uncommon thing to find the population migrating to where there is a queen, where they are accepted and augment the population. The two or three cell-carrying frames are then introduced with the trepanned or grafted larvae, the frames themselves well coated with honey. I endeavour to give these frames without undue disturbance. When removing the inner cover or quilt, it is frequently necessary to break the great mass of bees clustering from the cover, down the space between the dummy board and hive wall. These fall and break up on the floorboard. At this operation diluted honey is poured upon the frame tops. Whenever possible, this work should be carried out prior to any major nectar flow. It is most inconvenient to have your queen cells webbed together and the frames utilised for honey storage.

Many workers who use the system close the space between the dummy boards and hive walls to limit occurrences of this kind. I personally prefer to work with the huge hanging mass of bees, but occasionally inconvenience has occurred from an unexpected nectar flow. A suitable feeder with dilute honey is now placed in position. Systematic but brief inspection of the larvae will, after twenty-four hours, confirm acceptances. Ten days later, i.e., on the thirteenth day, the cell raising colony is taken by car to the mating apiary, where the finished queen cells are transferred to waiting nucleus colonies, it being understood that the colony is cleared of its own drones by screening beforehand.

After meeting Brother Adam of Buckfast Abbey I became so interested in the subject that I established my queen mating apiary on the Northumberland Fells. The splendid isolation of the area lent itself to this kind of project: and it was here that I did my work on "holding colour in the yellow banded bee". From its introduction into this country in 1859, all attempts to keep the yellow Italian bee yellow in the northern parts of Britain had failed, and many explanations were put forward. I finally established that if larvae from the selected yellow breeder queens were reared by yellow nurse bees colour was maintained without staining. But when the same larvae were given to colonies of local brown bees to rear, many of the resulting queens were brown, and staining was considerable. My trials extended over fifteen years, and during the whole of this time colour remained constant in the original Buckfast line. My work interested the Russians as their Biologists had had the same experience, and this began an extensive interchange of views on what we beekeepers term " the nursing factor." In 1958 I was invited to contribute to PCHELOVODSTVO, the Russian Bee Journal, and the article was illustrated with pictures taken at Shilford and on neighbouring fells.

I was pleased when someone followed up my reference to the "nursing factor" in the British Bee Journal, and Mr. Sandeman Alien of Halesworth, Suffolk, wrote: " If it is possible the whole edifice of queen rearing methods will be shaken." (Sandeman Alien had done useful work as Secretary of the Honey Producers'

Association, and his opinion was worthy of respect). The purpose of my notes was to convey to the general reader my difficulties in maintaining colour in yellow bees. As a queen breeder I had found that colour could only be kept constant in the North in yellow-banded bees by closely watching this point and by ensuring that larvae of the breeder queens were reared by yellow bees, and to this I attached a rider, that behaviour of the daughter progeny reflected on the colonies they (the queens) were reared in.

I am concerned with the influence of the nurse bees of a colony over larvae introduced to them from an outside source and I know this interests both Brother Adam and Dom. Leo Smith at Buckfast. Brother Adam says, "environment and nutrition in their larval state determine the worth of queens reared. In fact, an unfavourable environment can so far check the development of the hereditary make-up that certain characteristics will not even appear." Ruttner and Mackenson in their Genetics of the honey bee (Bee World 1952) make more pertinent observations. Reviewing work by Muller and Dreher, they say: "It must be borne in mind that with the bee as with other insects, the colouring of the exoskeleton may be affected by temperature, and hence the proportion of bees with yellow markings in a colony often fluctuates with the season."

The "nursing factor" and Mr. Sandeman Alien's letter to the British Bee Journal aroused great interest, but there has been a tendency to depart from my original implications. During the fifteen years I have had the "Buckfast bee" colour has only been maintained by selecting both queens and drones (for this factor) and taking particular care with mating. I would say that the colouring now is the same as when they first came here. The point I wished to make was this: when a batch of larvae from a yellow breeder queen happened to be given to colonies of local brown bees to rear, colour was lost overnight, and the virgins, instead of being yellow were brown.

I have often mentioned the difficulties Northerners experienced in maintaining colour in yellow bees, and how they were usually obliged to abandon the game when this darkening crept in; a point which has been confirmed by Mr. Hodgson Gray. But for those who persevere with a mating apiary, and are no longer handicapped with the problem of mis-mating, it is possible to get a constancy of factors (including colour) as I have proved. But can we be quite sure that we deal with pure races in our work? Until this has been established there is the possibility of mutations being sparked off at any time. My observations really interested the Russians, and Khalifman (the Stalin Prize Winner) whose book was so well received here, wrote to me about his own work on "The Vegetative Hybridisation of Bees." This correspondence was of the most friendly kind. You will gather that this gave some backing to the claims of Lysenko, who, in the first place, worked with plants!

Mr Khalifman, author of "Bees", Stalin Prize book, 1951.

The celebrated biologist has reappeared upon the Russian stage. My article "Bee Breeding" in the Russian Journal came at a time when Lysenko was re-establishing his reputation under the Kruschev administration. T. D. Lysenko's great work was with a rust-resisting wheat and he claimed that characteristics acquired during growth of the plant would be transmitted to succeeding generations. Lysenko's findings offended the accepted canons of heredity, and his work was generally unacceptable to the Western World. In 1959 Lysenko sent me this letter:—

"We have published for readers of the journal 'Agro-biologiya' which I edit, data about this observation of yours, which aroused great interest, not only among practical beekeepers but also as a general biological theory.

I think that if these facts about the formative role of rearing had been known to Charles Darwin he would not have been able to assert that the instance of nectariferous bees and other social insects presented the greatest puzzle in his doctrine of natural selection.

With best wishes,
(signed) Academician T. D. Lysenko."

Khalifman has replied to his critics in an extensive communication. The original in Russian, which runs to twenty-one pages is now in my possession. Several translations of the paper have been made. and one appeared in the pages of the British Bee Journal (May, 1959) and was hailed as being one of the most important ever to be published. In this we are brought face to face with Soviet work in the whole field of biology, with the possibilities of directed mutations in this sphere too.

For all who persevere with a mating apiary—whether in isolation on the fells—
or by utilising local features, such as the dense woods which abound in this part
of the country, headway can be made. Those who know my moorland mating
apiary will recall that the nuclei are terraced in a half-circle down a hillside with
the one or two drone supplying colonies high above them. The natural basin
escapes all winds, and while gales are blowing on the moor top, the basin is
a sun trap. The other important point is that the apiary is sufficiently isolated
amongst the hills to prevent any mating interference from an outside source. Late
Springs, are, however, a problem to the Northerner —at least I have found them
so—particularly when trying to over-winter a large number of young queens. The
small colonies stand up to the cold very well until March and April, and then start
to dwindle and go down fast. And apart from bringing them *inbye* to the apiaries
and setting them over a queen excluder on the honey-producing colonies, there
is little one can do.

I do this joining up during a cold snap, but no top entrances must be provided
until the small colonies are sufficiently boosted from below. I finally overcame
this problem by maintaining the colonies on seven British Standard combs
permanently on the moor, but certainly nothing less than six combs should be
used. The experiments which gave most satisfaction were those which involved
hand-picking drones, and placing these with about a cupful of screened bees,
and a virgin queen in a Swiss type of mating box with candy. These were set down
in the woods and queens were laying in no time. This enabled me to collect
such useful information in one season; in fact, actual turnover of queens is only
surpassed by Instrumental Insemination. In the mating apiary proper, where
nuclei are permanent and over-wintered, progress is from year to year.

The transactions of the Royal Entomological Society for 1956 contained Lord
Rothschild's paper on "The Spermatozoa of the Honey Bee" and Woyke and
Ruttner bring this up to date in January, "Bee World" of that year.

The position of queen and drone during copulation comes under discussion
and three illustrations are given. Bishop's (1920) face-to-face position, and two
alternatives of Herrod-Hempsall's (1937) drone mounts on back of queen.

I can throw some light on the latter work. The late Leonard Illingworth, who
was associated with the Apis Club, renowned for his work for international
beekeeping, and an impeccable worker and observer, told me how William Herrod-
Hempsall brought a paired queen and drone over to Foxton for examination prior
to preparing the photographic account for the second volume of "Bee-keeping
New and Old," and he (Illingworth) was left without any shadow of doubt that the
drone honeybee mounts the queen to mate.

My own "path observation" (May 1949) supported this, and my impression was one which might compare with mating Bumble Bees. The drone certainly joined the queen after she had made a long sweeping flight above the ground and in this position they drifted quietly until they came to rest. From subsequent observation of virgin nights, I am inclined to support the view that queen and drone are only attracted to one another in flight, and actual attachment occurs in the short melee, soon after they ground. The Swiss worker Fyg examines the whole question in "Bee World" for August 1952.

For years it was held that under normal conditions the queen honey bee only mated once, although reports appeared from time to time indicating that a second and subsequent matings did in fact occur. The accepted canons were eventually exploded and the whole subject was viewed from a different angle. Nucleus colonies were set up in the early summer of 1953 in the craters of Vulcano Island off Sicily, and these were under continuous observation until the experiments were concluded. The entrances to the small hives being closed with excluder zinc, the flight of almost every virgin was recorded. It was established that many queens mate several times during one day—the mating sign usually being removed a few minutes before (or after) the queen had returned to her hive. Queens which did mate once invariably were failures with a tendency to be superseded. The investigators agreed with the Russian worker Tryasko that queens mate with four or five drones per flight, and remove the mating sign themselves outside the hive. Sometime later the young Polish worker. Woyke corroborated this by sperm counts.

I find it hard to believe that loss of the mating sign is the rule, and we must consider the evidence put forward by the monks of Buckfast Abbey, which Father Leo Smith confirmed when he visited me in 1955. He has accompanied Brother Adam on his visits to the mating apiary on the wilds of Dartmoor: and when really early in the forenoon 8/9 a.m. before the bees in this moorland apiary are flying, they have often found a large percentage of the virgin queens carrying the mating sign from, presumably, the day before. This in itself, indicates that the sign is retained for longer periods than supposed by Tryasko: and I have had similar experiences in Northumberland. Looking back over the 1950's I feel that we should have more information about the strains and lines of bees used in these experiments. Were they, for instance, closely inbred for generations, and what about their nurture ? One report indicated that the virgins used passed through queen excluder zinc with ease, and were certainly much smaller than virgins we are accustomed to. And again in 1954, one of the wettest years we have experienced, I was quite sure that a batch of sixty virgin queens mated during one-quarter-of-an-hour flight, and went on to give a good account of themselves in 1955 and later.

In many of the more remote parts of the country, inbreeding has become probably the most pressing problem of our time. Brood viability is low and many colonies exist where brood dies at certain times of the year. Addled brood appears to be the convenient term by which to describe this.

Some years ago Mr. Hayden (Sunderland) told me that he accidentally damaged a queen in heavy lay—the vent or something was nipped—and she was unable to pass an egg. He was most surprised, and I suspect shaken, to find upon his next inspection that the egg had duly hatched while still being carried by the queen. This gives additional backing to Herrod- Hempsall's claim that the eggs of the honeybee could hatch without food being placed in the cell.

In my round of the apiaries, I occasionally come across colonies with food chambers throwing out larvae. So in one particular case in the 1940's I opened up and found a drone-laying queen. What a mess! Every conceivable cell was filled with eggs and drones were everywhere. But the interesting thing was a large part of them were albinos, and I know of no more fascinating sight than finding a colony with a white-eyed drone population. Harry Thompson who did so much to pioneer the instrumental insemination of queen bees would have rejoiced in this particular find.

On another occasion I received a queen from Wooler with a heavy infestation of Braula coeca. For many years the late William Herrod-Hempsall said that he had never found the bee louse living North of the Humber, and presumed it could not withstand our winters. But in April 1949 the great man himself was staying with us for a lecture tour of Northumberland, and when we went North to Wooler, a beekeeper from the same district brought a queen (with Braula) to the meeting, and we subsequently removed from her eighteen healthy specimens. This indicates that Braula is found all over the British Isles as Miss Logan has mentioned it in Ross-shire.

The late Jim Cox of Berks, often mentioned environment and incubators. I still have one in the bee-house, though it is no longer used. Apart from purely experimental work an incubator should have no place in queen rearing. Queens hatched in an incubator were always inferior to queens which emerged naturally in a bee colony. This showed up most noticeably in the brood pattern, in the weeks prior to going to the fell—and usually ended in supersedure. The other thing which bothered me was the fact that you had always to be about when the queens hatched to remove queen cells from the cages, otherwise the virgin would re-enter her cell head first, the lid would close, and thus imprisoned, she would die. On other occasions, I placed pieces of comb containing eggs in the incubator to test Herrod-Hempsall's claim that eggs would hatch without food being placed in the cell. But this experiment failed.

In 1957 there was an almost complete absence of drones in all colonies here (both adult and in brood stage) and there is remarkable unanimity of opinion about this amongst beekeepers I have met. For a long time I have wondered if there is any link between the pollen gathering activities of honeybees and the weather. The late Bob Robson always held that when the bees worked the buttercups and there was a spate of pollen in his apiaries in April/May this presaged a poor summer, and my own observations tend to bear this out. In May 1954 there was the heaviest intake of Dandelion pollen I have recorded here, and a notorious summer followed. Similarly, in 1958, about as much pollen went into the hives and we endured another very trying season. In both cases, the amount of pollen about struck one very forcibly. When handling combs from six-comb nuclei to double brood chamber lots, bees were at a loss what to do with it.

1 Khalifman (personal communication) has described the part played by honey bees in growing quality grass. For long bees have been known to visit grass, presumably for moisture. But work in Russia indicates that they carry a yeast capable of increasing protein in the plant. Farmers who have grassland within the foraging areas of bee colonies benefit immeasurably.

" *When her offspring do not swarm*
When inspected do not harm,
You have got a queen of gold
Never leave her to grow old."

A.R.C. — Bee Craft.

CHAPTER NINE
PROBLEMS

" The sick, for air, before the portal gasp.
'Their feeble legs within each others clasp ".

VIRGIL.

ONE piece of good fortune has befallen those people whose lot it is to keep bees in this part of the British Isles, namely, the comparative absence of major bee diseases within the area.

Brood diseases have presented no alarming problems to date. European Foul Brood has only occasionally shown up; this was in the wet years of 1927, 1946 and 1954; and we appear to be fairly clear of American Foul Brood too. The late Bob Robson of Wooler (not to be confused with his namesake at Riding Mill) who was a former secretary of the old Northumberland Beekeepers' Association, always held that the rather dry conditions in the North East had something to do with it.

But against this we have had our share of adult bee complaints. Isle of Wight disease (later to be known as Acarine Infestation) wrought havoc in the years immediately preceding the first World War (1914/1918). The native honey bee population was decimated to such an extent that the County Beekeepers' Association was disbanded in 1917, and was not reformed until 1936 (although an active group of beekeepers. namely Bob Robson, Riding Mill, and John Howstan, Newcastle, got an association going at Newcastle in 1922). Robson and Cessford gave me many graphic accounts of the epidemic, as they remembered it, and how they combated it at Riding Mill.

Often in mid-summer, when the clover flow was at its best, bees just poured out of their hives—as if preparing to swarm— leaving both brood and honey behind. They covered the ground in one great crawling mass. They tried everything from painting the insides of the hives with creosote to the remedies of Samuel Simmins and Ayles. Apparently at that time "Ayles Cure" was quite the thing until Frow came into the picture. In the hot summer of 1921 they received some golden queens from Jack Tickell of Cheltenham, and they liberated these directly onto what brood was left within the hives. It seems that all queens made headway, for there were soon some flourishing small colonies in the apiary. Of course conditions must have been ideal. But this provides a good illustration of the perseverance and endeavours of these old Northumbrian craftsmen.

Acarine infestation, as we know it, in no way compares with the description of those early days, and in all the cases I have run into here there has been a good deal of Nosema about, too, I was particularly interested in Dr. Jeffree's paper (Bee World, January 1959) listing probable acarine "areas" of the world, as he once stayed with us at Shilford.

Nosema disease provides what is probably the biggest headache in Northern beekeeping today—and as yet there has been no serious attempt to come to grips with it. Without any doubt there is a distinct tie-up with heather-going conditions and the complaint, and it is likely there are countless brood combs in use which have at some time in their long life been contaminated with Nosema spores. It is still an accepted thing in some areas to expect some colonies to go down during the winter, and upon examination in the spring, the familiar staining is usually in evidence all over the hives.

These combs, providing they are not too badly holed by mice, are cleaned up and held for future use, and so the cycle goes on. Brood rearing sooner or later takes place in them when they are brought out, and winter stores in the form of honey from the ling eventually fills the cells. The colony housed on such combs in turn becomes infected during the summer and goes down, or, if it does manage to pull through, it is in a very bad way indeed. I have always found that bees recently imported from Italy and the U.S.A. are particularly prone to Nosema, but that the resistance of the local lines of bees is good (I am referring to the brown bees of the area). My original yellow Buckfast line has been always clear of the malady, and the point of interest here is that many of my colonies are wintered in moorland areas on stores made up almost entirely of heather honey.

During the 1950's the Ministry of Agriculture, Fisheries and Food launched the Apiary Health Certification Scheme for Bee Breeders and for the first time beekeepers throughout the British Isles were able to obtain home-bred bees and queens, from apiaries free from American and European Foul Brood and Acarine, Amoeba, and Nosema diseases.

Tom Seckstone for many years apiarist to the late William Herrod-Hempsall was the first bee-keeper to hold the certificate. But through the untiring efforts of the Ministry's Chief Bee Officer, Mr. P. S. Milne, several other apiaries have been inspected by officers of the Ministry, and had bee samples, etc., examined at Rothamsted. A certificate (No. 1. 60) was issued to the writer in the Spring of 1960.

(1) The late MR. J. N. KIDD (Stocksfield) imported bees in skeps from Holland at this time for re-stocking purposes, and he contributed to the early numbers of Bee World [1919].

(2) The author practices complete destruction. All infected material, bees and brood combs are destroyed and hive parts thoroughly washed. His efforts in fumigating combs, etc., and feeding Fumidil B. have been unsuccessful. Many brood combs are renewed annually in the apiaries (where there is no known infection) as a matter of routine.

"If any of your stocks are weak,
Be diligent the cause to seek.
Don't hesitate for help to ask
It's the expert's willing task."

A.R.C. — Bee Craft.

CHAPTER TEN
PEOPLE

" . . . as bees flee home wi' lades o' treasure.
the minutes winged their way wi' pleasure ".

ROBERT BURNS — "Tam O' Shanter"

THE ILLINGWORTHS

IT is often said that the indiscriminate and mis-directed use of insecticides and other toxic chemicals hangs like a curse over all beekeeping today. Richard's work in Norfolk, Whitcombe's in the U.S.A., and that of Lysenko and his colleagues in the U.S.S.R., has established absolutely the honeybees' role as a pollinating agent. In the early 1950's, the late Leonard Illingworth who was so closely associated with the Apis Club for so many years, was obliged to give up beekeeping in the Foxton district of Cambridgeshire owing to this spraying menace. He wrote me many letters in the evening of his days, describing how he, and his devoted sister Mabel, went down daily in the early summer months to their hives to shovel up the dead and dying bees!! Colonies which had come through the winter strong in bees (Illingworth with his Langstroth hives was a great believer in the food chamber) were decimated overnight, only the brood being left in some cases. Fisons have recently (July 1959) made a bold approach to tackling this problem. Representatives of the beekeeping organisations of this country met their representatives at Chesterford Park, Essex, and the whole problem was thrashed out. The outcome of this meeting is that there is likely to be greater co-operation between beekeepers and growers in the years to come.

left, Leonard Illingworth, 1951. Right, lecturing.

The Illingworths lived in a lovely old-world thatched cottage called "The Way's End," and once when attending the British Cattle Breeders' Conference in Cambridge, I visited them there. The family were of Yorkshire extraction, and Leonard, who was the youngest, was far from strong. Teaching was his vocation, and for some years he was associated with Winchester. For health reasons he repaired to a rural area of Cambridgeshire, and his sister Mabel, in due course, joined him there and mothered him ever afterwards. Just as he was fine, and small in stature, his sister was robust and strong, and her great love was attending to the orchard and the geese.

Illingworth throughout his life favoured Langstroth hives which he operated on a brood chamber and shallow super system of management. He ran sixty colonies in one apiary, and for years (until spraying made it impossible for him to keep bees any longer) took his main crop from the sanfoin. He was keenly interested in queen rearing, and, being a prolific writer, he was soon in touch with the Swiss workers, whose methods he publicised here. In 1936, when Miss Betts found the joint offices as Editor of Bee World and Secretary of the Apis Club altogether too much, Leonard Illingworth relieved her of the latter, and consolidated the destinies of the club with skill. He also found time to be Secretary of Bee Diseases Insurance. A well-known and respected figure at International Congresses, he was eagerly sought after as a lecturer in this country. He visited us often in the 1940's, and spoke to many gatherings of visiting beekeepers. We always found him a delightful companion; an original and dedicated man. A conversationalist of repute he managed to get a continuity of sentences with the frequent use of "and-er". In the summer months he always wore a boater: and when he was with us, worshipped regularly in the little village church at Riding Mill. When he was not beekeeping or writing, Illingworth's other great love was chess, and the Hastings Chess Congresses were religiously attended every year and were a source of great delight to him. Although in later years his eyesight handicapped him greatly, the British Beekeepers' Association managed to make him its President in recognition for his work for International beekeeping just before he died.

As Secretary of the Apis Club he was responsible for the library started by Dr. Abushady. This was housed in the dining room at the time of my visit, and I have never seen a room so full of books!! The Bee World's as they came in monthly from the printer were all hand-wrapped at Foxton, and carried by brother and sister in a washing basket to the Post Office.

But my own memories of this pious, God-fearing man, are centred around the thatched cottage, and the two cats which he tolerated to sit one upon each shoulder.

THE HERROD-HEMPSALLS

Probably the most colourful figures in British bee-keeping were the Herrod-Hempsall brothers, who followed Mr. T. W. Cowan[1] and Mr. W. B. Carr as leaders of the craft as the Victorian era drew to a close. It fell to William Herrod-Hempsall to bridge the gap from skep beekeeping to the present day, and to evolve a system of management for movable comb hives, which would be acceptable to the country as a whole. Having once made up his mind he adhered fervently to the one course for the next fifty years. It is unlikely that this country will ever produce another craftsman of the calibre of William. From humble origins in Nottinghamshire, his tremendous enthusiasm and gifts soon attracted the attention of the celebrated Mr. Cowan, who had fathered beekeeping here from almost the inception of the British Bee Journal in 1873, and British Beekeepers' Association a year later. He was soon invited to come to London, where, once established at the "Bee Journal" office, he became in turn Secretary of the British Beekeepers' Association, Editor of the "Bee Journal" and the most powerful force in beekeeping for the next forty years. When Cowan, through old age, retired from the scene, William's brother Joseph came South to become Editor of the paper, while William himself kept a tight hold on it financially.

Herrod-Hempsall, on the left of the second group, in his usual country clothes.
L-R B. Leafe, Mr Wright, Mrs Lowsley, W. Hamilton, W. Slinger,
G. Foster (Chairman YBKA), Miss H E Dickenson,
Lord Huntingdon Parliamentary Secretary (House of Lords), Mr t. Williams,
W. Herrod-Hempsall, J. S. Rowntree, H. Molyneux, Mrs Ivy Jacques & F. Lucas.

During the great Isle of Wight scourge, he was charged by the Government of the day to visit Italy to arrange a large scale importation of Italian queens bees into the country, and this work more than any other, was responsible for re-establishing beekeeping in Britain after the first World War. About this time, on the advice of Mr. Cowan, William was installed as technical adviser on beekeeping to the then Ministry of Agriculture and Fisheries, and in this capacity he was, apparently, free to interpret the duties as he alone thought fit. But he at once realised its possibilities. He set out in the succeeding years to visit personally almost every beekeeper in Britain worthy of the name. He lectured and gave demonstrations in every corner of the country, and the name of Herrod-Hempsall was soon a household word wherever bees were kept. He wore out cars galore touring the country, and we all wondered how he kept the pace. For all this he was rewarded with the magnificent sum of £500 a year, but he eked things out by selling books (which he somehow found time to write amid his journeying). buying and selling honey, and by sleeping most summer nights in his car. But I imagine Brother Joe helped him quite a lot in London. His "Beekeepers' Guide" which followed Cowan's book, needs no mention here, as almost every beekeeper has had a copy through his hands.

But we must look at his life's work, the 1,800 page two-volume tome which he entitled "Bee-keeping New and Old." It has always been a source of wonder to me that any one person could set out and complete a work this size and do so many other things as well. I first saw the volumes at sixteen: and no beekeeping books before or since have so fired my imagination. It is only natural that the information they contain will be amended by succeeding generations in the light of current knowledge. But this in no way detracts from the fact that they are books of classic import. Few works are more profusely illustrated, and this says much for William's skill as a photographer. Every beekeeper should aim to see the volumes sometime in his life.

When, in later years, William stayed with us at Shilford, he told me that the chapter which had given him most pleasure to write was that on Heather Honey: and it is obvious that the author relished his heather-going experiences with the Northern beekeepers. He established a large apiary of about 200 colonies, housed in all double-walled hives, and a school of beekeeping at his home at Luton, Beds., and Tom Seckstone, a most able craftsman in his own right, was responsible for it for a great number of years. Personalities from all over the world visited the apiary, and it was the scene of several great field days. The most notable was Dr. E. F. Phillips, the American beekeeping leader, when he visited this country on Apis Club business in 1926. This meeting was very well publicised at the time, but in later years William cared not to be reminded of it. With the great upsurge of popularity in bees and beekeeping during World War II, William and his brother decided to retire from public life for the time being (but Joe still produced the "Bee Journal") to let the newcomers have their head as he was convinced that he would be called back within a year or two. He never lived to see that day as he passed away in September 1951 at the age of seventy-nine, but he managed to see things swinging back in his direction. William Herrod-Hempsall wrote me many interesting letters during this period which I have kept, as they give some indication of his forceful personality. When he retired from the Ministry in 1944, a testimonial fund was launched, and the amazing sum of £1,550 was contributed by the beekeepers of this country: people whom he had visited and helped over the years.

His last lecture tour of Northumberland was in April 1949 when he made his headquarters here. It was an Easter week, the weather was a treat, and this was the first sample of the heat-wave which was to last the summer. He was a sound philosopher, but he could not tolerate smart Alec's and dealt with them accordingly, and on this score alone was often misunderstood. After all that has been said about William, I am extremely glad to have known him, and I have made my own evaluation of his work.

April I0th, I949.

My Dear Colin,

    I am sorry that I have been so busy that any writing has been taboo, hence the delay in sending the enclosed.
    I arrived home quite safely, and in comfort as the train was not crowded.
    Many many thanks for your kindness to me and the splendid arrangements you made for my comfort to and from the centres where the lectures were held.
    I congratulate you on your bee-keeping , also your work for the association which I know from experience is arduous and absorbs much time from your private life. It was quite evident to me that your effort and energy has entirely rejuvinated bee-keeping in Northumberland, and I wish you every success in the future both in your apiary and association, not forgetting your splendid dairy herd.
    A word in your ear , do not let the communists of bee-keeping in London get a hold on your association and muck up all the good work you have , and are, doing.
    If at any time I can assist you with advice, or in any other way,please command and it will be a pleasure to obey.
    Please give my kind regards and good wishes to Bob Robson.
              With kindest regards and hearty good wishes now and always.
                Yours very sincerely,

W. Herrod-Hempsall

Letter from W. H-H to C. W. after his last lecture tour of Northumberland.

(1) The author has followed the advice of MR. L. E. SNELGROVE here (who knew Cowan and has referred to the late T. W. Cowan as " Mr." We understand that at least one degree was awarded by an American University honoris cmisa (but apart from written work the degrees were never used by Mr. Cowan in his lifetime).

BROTHER ADAM

I have already mentioned my correspondence and subsequent visit to Brother Adam, and the Benedictine Monks of Buckfast Abbey.

Charles Kherle, who became Brother Adam, came to England from Germany at the age of eleven, and he at once began to assist Brother Columban in the Buckfast kitchens. The monks had already begun their lifes' work raising the Abbey Church of St. Mary in the lovely valley of the Dart. Brother Adam in time helped Brother Columban to tend the few colonies of bees he had established there: his interest was kindled, and within a short time he had a model apiary on its feet. At this point Brother Columban fades out of the picture, but not before his name was indelibly linked with a recipe for candy with which at times he fed his bees. This was given to beekeepers through the medium of the "British Bee Journal."

In the succeeding years the Abbey apiary was steadily increased in size.

At the outbreak of the first World War the German population at the Abbey were confined to the Abbey grounds—as there was some local hostility—and time hung on their hands, so they turned to bee-keeping. The apiaries were again extended, and as Italian queens became available they were made use of, and a system of management developed specifically to suit the larger colonies.

Brother Adam now embarked upon the ambitious bee-breeding programme with which his name will always be associated. An isolated site was found on the wilds of Dartmoor plateau, and a queen mating apiary was soon brought into use.

Around 1930 the double-walled type of hive and double brood chamber with British Standard frames were abandoned in favour of a single-wall type taking twelve Dadant frames which could be worked on a single brood chamber system of management.

At the time of the first World War a large number of colonies were kept in one apiary at the Abbey, but then out-apiaries were established under various conditions in the Devon countryside holding from 35/40 colonies each, and 400 or more colonies were accommodated in ten apiaries.

These have been chosen with great care, and when I have visited Brother Adam during the summer months, and he and his team have been working in the out-apiaries, I have been struck by the same methodical lay-out as in the home apiary itself. Almost every out-apiary has its own bee-house.

(a) Brother Adam
(b) Br Adam at Colin's moorland apiary
(c) Brother Adam and Colin
(d) Br Adam on his 95th birthday at Colin's
    farmhouse
(e) Peter Donovan, who gave enormous
    help to Brother Adam during his work

In the Buckfast apiaries all the colonies are grouped in fours, with hive entrances facing in different directions, instead of in the traditional straight lines. This grouping was adopted at the heather in the first instance to cut down drifting, and I introduced it at Shilford in due course.

I have some very happy recollections of my visits to Brother Adam, particularly the occasion when he took me out to the mating apiary on Dartmoor. The steep, narrow, wooded lanes as you ascend the moor from the Dart valley really take some negotiating. I was driving and I think my host was very tolerant when I had to change down into bottom gear half-way up the hills: but the 400/500 nuclei in the special type of mating hive are certainly worth going a long way to see.

Brother Adam now holds a unique position in the world of beekeeping. Nevertheless, he is, at heart, a very humble man, dedicated to his life's work. I received a letter from him at Christmas, 1956, which expresses more adequately than I can ever hope to do the real sincerity of the man.

DR. EVA CRANE

Soon after the war, the British Bee-keepers' Association decided to set up a Research Committee with my old friend Graham Burtt of Gloucester as Secretary. Shortly afterwards, Dr. Eva Crane of Hull came into the picture. A Doctor of Philosophy, she was unusually well suited to make a contribution in this field. Dr. Crane's flare for this kind of work was soon to make itself felt.

A young Eva Crane lecturing at Riding Mill, July 1949.

Eva Crane followed Graham Burtt as Secretary of the Research Committee, and by the end of 1948 there were indications that there was sufficient interest in the country to justify the setting up of a separate association concerning itself solely with this particular aspect of the craft. Thus, the Bee Research Association was launched in the Spring of 1949. Everyone appreciates what Dr. Crane has done for British and world beekeeping since then.

With the inauguration of the Bee Research Association the affairs of the Apis Club (founded in 1919) were wound up, and Dr. Crane followed Miss Betts as Editor of "Bee World" in January, 1950. I never met Miss Betts, although she was pointed out to me at one of the National Honey Shows in London.

Dr. Crane came North in the summer of 1949 to lecture to a large gathering of beekeepers at Robson and Cessford's celebrated apiary at Riding Mill, and, in fact, this was the last great meeting to be held there before their life's work was disbanded. Beekeepers are indeed fortunate to have such a talented person in their midst.

MR. JAMES CUNNINGHAM

One of the people who gave me tremendous encouragement in my teenage years was James Cunningham, Editor of the "Scottish Beekeeper" and for many years lecturer and adviser to the Edinburgh and East of Scotland Agricultural College. I was still fairly young when I set out by train for Edinburgh to take the Bee-Master Examination in the college apiary. Mr. Cunningham was my Examiner, and I was set to work upon a Glen Hive (the like of which I had never seen before) but somehow or other I got through this test, and returned home much elated. James Cunningham then came to Shilford quite a lot and he once had the novel experience of acting as invigilator for my British Bee-keepers' Association Senior Examination (Part II, written).

MR. ANDREW LIMOND

The following Autumn, 1946, I went North again to Glasgow, where that much-beloved figure in Scottish bee-keeping, Mr. Andrew Limond, of Maybole, Ayrshire, put me through the practical part of the Honey Judge Examination in the Kelvin Hall, the written paper having been completed sometime before. The experience at Glasgow brought me into contact with the Scottish beekeepers for the first time, and was certainly worthwhile. When judging at the Scottish National Honey Show (October, 1959) I visited the Old Mill at Minishant and saw the apiary immortalised by Andrew in the "Scottish Beekeeper". The Limonds always recall with pleasure that happy time in Glasgow's Kelvin Hall.

MR. A. R. CUMMING

After I had had bees for a few years I became more and more impressed with the work of the North of Scotland Agricultural College which encouraged the crofters in the Highland areas to keep up to forty colonies of bees. Where this was put into practice the sale of heather honey to tourists augmented many humble incomes. Mr. A. R. Cumming, formerly of Kilmarnock, who retired to Inverness, wrote me many letters during this period. Just after the war, he produced what was probably up to this time, the best colour film to be made in this country, "Bee-keeping in the North." Shot mostly in the Highlands, the colour photography was of a high standard. It was first shown in England at Riding Mill in April 1947, when over three hundred people crammed the village hall. I well remember my difficulties that day. Cumming undertook to despatch the spools by express post from Inverness the day before, but by late afternoon they had not turned up despite frantic calls to a sympathetic G.P.O. in Newcastle: but at last, at 5.30 p.m. they were located. A volunteer motored into Newcastle, claimed the prizes, and the great show went on at Riding Mill without a hitch at 7 p.m.

Mr. Cumming during his lifetime collaborated with one of Scotland's greatest personalities, Margaret Logan of Ross-shire, to produce that land's own beekeeping classic, a book called "Bee-keeping Craft and Hobby." This joint effort has provided a never-ending source of pleasure to all who ever come to take it up.

A number of years were to go by until one day Nancy Young, who was at one time Editor of the North of England's own beepaper "Northern Apiarist" brought Margaret Logan out to tea, and another friendship was established.

MISS LOGAN

Margaret Logan (or Peggy, as she is known amongst her friends) and her sister Nancy were protegees of that renowned Scottish Bee-Master, Dr. John Anderson. Peggy Logan followed her sister to what is probably the largest area embraced by any college authority in Europe, namely, all the Highlands and the Islands in the Northern half of Scotland.

Miss Logan has carried on the grand traditions of her mentor, and still takes her stand by the great Glen Hive; she is, in fact, one of the few people now left who can use this hive successfully.

Margaret Logan is altogether a fascinating personality, and she, too, has a very special place in British Beekeeping.

MR. R. H. SKILLING

The Scottish Beekeepers' Association has been extremely fortunate in finding good Editors for its official organ "Scottish Beekeeper."

Both Dr. John Anderson and Mr. James Cunningham made notable contributions when occupying the Editorial chair—and in recent years Mr. Robert Skilling, an immensely popular figure in northern bee-keeping circles, has guided its destinies with dexterity and skill. Appearing monthly, the "Scottish Beekeeper" deserves a very wide circulation.

R H Skilling

DR. MALCOLM FRASER

Over the years Dr. H. Malcolm Fraser (Middlesex) has made an immense and enduring contribution to the cause of British beekeeping. His two books "Bee-keeping in Antiquity" and "History of British Bee-keeping" are in a class quite by themselves.

For thirty or more years he has been a regular contributor to the "British Bee Journal" and his column "Regurgitations" has become one of its most widely read features. Striking in appearance and benevolent in attitude, Dr. Fraser is one of the people you look for at the Royal Agricultural Society's Show and the London National Honey Show. Invariably he will be found surrounded by a party of entranced schoolchildren as he describes with charm the mysteries of the hive, and what makes this feat all the more amazing is that Fraser's immense age has never dimmed his enthusiasm.

The Royal Show revisited Newcastle in 1956 and Dr. Fraser came North. I went with my colleague Mr. Joe Smith, who is associated with the training centre at Ainwick Castle—and incidentally a person who has done much for beekeeping in North Northumberland—and we entertained Fraser to lunch and he delighted us with anecdotes.

MR. H. J. WADEY

When lecturing at the Merrist Wood Farm School, Surrey, in July 1958, I met Mr. Jim Wadey, Editor of "Bee Craft", for the first time. I had read his books and followed his work with keen interest over the years. Wadey has made his own contributions to beekeeping and endowed with great enthusiasm he has a special place in our affections. His writings have a conciseness all of their own, and deserve to be widely read. I find his "Behaviour of Bees and Beekeepers" a treat.

JB Jim Wadey, 1958

H. J. Wadey wrote enthusiastically about beekeeping and added humour to "Beehaviour of Bees and Beekeepers" by commissioning Neil Nettleton to illustrate the book with cartoons.

MR. C. C. TONSLEY

From 1950 onwards new names began to appear in British beekeeping circles and amongst them those of Cecil and Nora Tonsley. Cecil Tonsley joined Joe Herrod-Hempsall at the "Bee Journal" office and eventually succeeded him in the Editorial chair. From 1954/1960 he was Secretary of the British Bee-keepers' Association. Under the Tonsleys the "British Bee Journal" has regained much of its prestige and this has only been achieved by real determination and hard work. Cecil Tonsley in comparatively few years has become one of our national figures, and the "Bee Journal" offices the rendezvous for almost every beekeeper visiting London.

(a) Cecil Tonsley at Shilford, May 7th, 1961. (b) Mr and Mrs Dermatopoulos (from Greece) with Cecil and Norah Tonsley at the BBJ offices at Geddington, Northants, July, 1982. (c). The editorial room with the Herrod-Hempsall and Cowan library at the Tonsley's home at Whitegates, Geddington, Northants. September, 19th 1992.

MR. F. AUSTIN HYDE

We turn now to the North Country—Yorkshire—and that lovely district to the south of the Cleveland hills which embraces Farndale, Helmsley, and Pickering, and villages like Sinnington.

It fell to Mr. Austin Hyde when Headmaster of the Lady Lumley Grammar School, Pickering, to make in the early years of World War II a direct approach to the then Prime Minister, Mr. Winston Churchill, for an allocation of sugar for the bees. That Churchill found time to go into the matter is shown in Volume IV of his War Memoirs "The Hinge of Fate" (page 849) and beekeepers in all parts of the United Kingdom took advantage of this in the years to come.

As broadcaster, lecturer and playwright, Austin Hyde personifies the North country, and his play "Home Cured" is particularly well known. In Mrs. Hyde he has a gracious beekeeping supporter, and it says much for their prowess on the Show bench that their exhibits at our leading Honey Shows invariably carry red tickets. It is not too much to say that Austin Hyde typifies the description "A great Yorkshireman."

CHAPTER 11
HEATHER HONEY

"Wild blossoms of the moorland, ye are very dear to me;
Ye lure my dreaming memory as clover does the bee;
Ye bring back all my childhood loved, when freedom, joy, and health,
Had never thought of wearing chains to fetter fame and wealth."

Eliza Cook

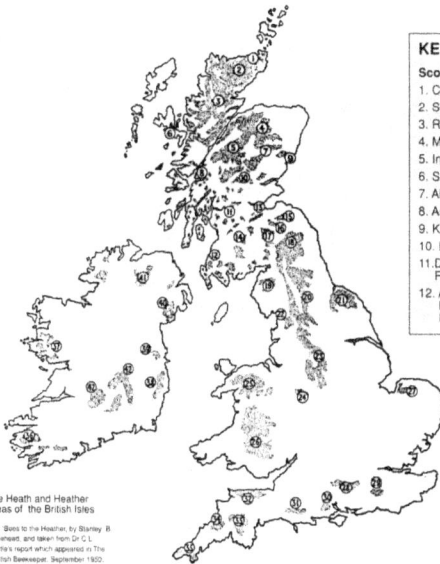

KEY TO MAP

Scotland
1. Caithness
2. Sutherland
3. Ross-shire
4. Moray-Nairn
5. Inverness-shire
6. Skye and Raasay
7. Aberdeenshire and Banffshire
8. Argyllshire
9. Kincardineshire
10. Perthshire
11. Dumbartonshire, Renfrewshire, Lanarkshire
12. Ayreshire, Wigtownshire, Kirkcudbrightshire, Dumfriesshire
13. Pentland Hills
14. Lanarkshire
15. Lammermuir Hills
16. Moorfoot Hills
17. Selkirkshire
18. Cheviot Hills.

England and Wales
19. Cumberland, Westmorland
20. The Pennines
21. Bowland Forest
22. Cleveland Hills, York Moors
23. Derbyshire
24. Cannock Chase
25. North Wales
26. South Wales
27. Norfolk
28. Hampshire, Surrey
29. Sussex
30. Hampshire
31. Dorset Heights
32. Exmoor
33. Dartmoor
34. Bodmin Moor
35. South Cornwall

Ireland
36. Mountains of Kerry
37. Connemara
38. Wicklow Mountains
39. Dublin Hills
40. Mourne Mountains
41. Sperrin Mountains
42. Slieve Bloom Mountain

The Heath and Heather
Areas of the British Isles

from 'Bees to the Heather, by Stanley B.
Whitehead, and taken from Dr C L
Whittle's report which appeared in The
Scottish Beekeeper, September 1930.

Bomber pilots with time on their hands at the end of the war made aerial photographs of Britain's moorlands, particularly in Scotland. At that time 49% of moors were being managed for grouse. By 1980, 57 sites remained as active grouse moors whilst 46 were no longer under management. Over those 40 years 24% of heather cover was lost on the managed grouse moors and on those where no grouse shooting occurred, the reduction in heather cover was 41%. Managed grouse moors provide an important habitat for many birds, animals and insects, yet very little is being done by the government to protect these vital environments, which are sadly decreasing year by year. For a commentary on this important issue, which affects beekeepers, readers would be advised to get hold of a copy of "Nature's Gain: How game bird management has influenced wildlife conservation" which can be downloaded as a pdf from the Game Conservancy Trust's website: www.gct.org.uk

Colin Weightman (centre), with Brother Adam on the right. Brother Adam's elequent description of Heather Honey made all those years ago to the 300 beekeepers who assembled to hear him at Newcastle in December 1934: " Red brown, like the water of the peat bog. A Gift of Nature carrying the tang of moorland air." still applies today.

Seek Permission before Placing Hives on the Moors

We are dealing in this article with areas [1] in the British Isles where the ling heather Calluna vulgaris is found and, to a lesser extent, the bell heathers Erica cinerea & Erica tetralix. The Erica spp provide Scottish beekeepers with a pleasing run honey - usually in July - which can be readily extracted. For the beekeeper, the best heather moors [21] are the well-managed grouse moors, where the old ling is burnt on a regular basis to encourage new growth, therefore the goodwill of Land Owners, Farmers and Gamekeepers must be sought and their requirements respected at all times. It is an unfortunate fact that many members of the public demand that they should have unrestricted access to open moorland and hillsides - often causing havoc among wild life and nesting birds. Beekeepers, too, often in their early days, unwittingly fall into this trap by moving their bee colonies on to the moors in late July and August and setting the hives down 'willy nilly' without prior consultation with anyone. Hives of bees, from which supers of blossom honey have been recently removed and which can have unusually vindictive house bees, are sometimes set down behind walls, close to footpaths, bridle paths and shooting butts, causing annoyance to walkers, pony treckers and shooting parties and their beaters, Before setting hives down anywhere, prior arrangement regarding permission, rent and siting, must be obtained from the parties concerned. Failure to do so can sometimes lead to the confiscation and loss of hives.

Arrangement of Hives

A heather stance [2] must be chosen with care - often a gully, or sheltered valley, can be found where the hives can be set down on their own. Make sure the spot will not flood as [3] there have been many occasions when hives have been partly submerged and even washed away after torrential downpours on the moorland tops. Shelter is of the utmost importance and the bees should be able to fly out up the hill and come down hill loaded. Often, dark clouds of homecoming bees can be seen skimming the wall tops as they return to the hives after battling against strong winds on the open moor. Avoid communal stances where anything up to 100 stocks of bees are set down together - often in straight rows - as there is a tendency for the bees to drift. A group of hives with the entrances facing in different directions is a better arrangement altogether. In small-scale operations, a simple hive stand carrying two hives, which can be carried on a pickup when moving bees, is a useful additional piece of equipment to enable the hives to be kept off the ground, However, many stocks taken to the heather are set down on the ground and levelled up with small pieces of stone. Unless a temporary alighting board is placed in front of the hive, growing heather or bracken may soon restrict hive entrances, compelling the returning foragers to run down the hive fronts to gain entry. Entrance blocks reducing hive entrances to around 12 inches in width and 3/4 inch high should always be in use to discourage slugs and rodents from gaining entry. In hot summers adders are sometimes found under the hives when they are placed directly on the ground and rubber boots are best worn to avoid snake bites or the bees from attacking the ankles. For this reason, too, it is a good idea to wear industrial or strong leather gloves when lifting hives off the ground.

A protected heather stance on Glen Clova, photo Beowulf Cooper.

It is another unfortunate fact of human nature that whenever a beekeeper finds a stance where over the years bees do well, other beekeepers will endeavor to take their stocks there too - and these moorland apiaries become eventually overstocked with bee colonies.

Twenty colonies to a stance - not far apart - will, if the colonies have been properly prepared, result in consistent performance - usually a super of sealed heather, comb honey, per hive as against a lot of unsealed honey where too many colonies are set down together. Some beekeepers errect permanent stands on the moors to carry the hives and use a spirit level to ensure that the combs hang correctly in the supers.

Preparing the Supers

For really choice combs of heather honey, full sheets of medium thickness worker brood foundation (unwired) should be used. However, many heather-going beekeepers use 'starters' only in the supers from cutting a sheet of shallow worker brood foundation into three strips. When using this method the top part of the comb is usually drawn out with the attractive flat matt capping of worker cells while the rest of the comb has the drone cells which some customers prefer. The practice of using extra thin foundation should be discouraged as all too often this collapses in really strong colonies of bees and many fine combs are spoilt each year. Only unwired foundation is used in the supers for cut comb honey and pressed heather honey. When hives are placed on stands on the moors, occasionally during a really heavy nectar flow and when there is excessive crowding within the hive, the bees will come out of their hives and build wild comb which they fill with honey under the floorboards. This happened in 1911, 1949, 1955, 1969, 1972 and 1981. 1949 had what was probably the earliest and heaviest nectar flow from the ling on record. Heather honey was being stored in the combs by July 20th, and bees being worked on a small brood chamber of 8 - 10 British Standard combs filled and sealed a super a week: this went on for five weeks!

Many beekeepers prefer to use drone comb in the supers. If the honey is to be for cut comb, customers also prefer it too.

In late seasons it is an advantage to use drawn comb in the supers - combs drawn out earlier on the Oil Seed Rape can be extracted - and then given back to the bees above clearer boards for them to clean. If the extracted combs are sprayed with warm water they will make a very good job of it, and you will be left with some fine dry, drawn out combs from which all trace of the OSR honey has gone.

There must be no trace of OSR honey in the hives when you take them to the heather otherwise it will ruin your crop of heather comb honey as early crystallisation sets in. Combs of stores containing 'set' OSR honey must be removed, too, from the brood chambers and replaced with brood combs of heather honey - if you have them - from the season before. It is a sound investment to get some brood combs filled with heather honey when the bees are on the moors as such combs are wonderful for feeding purposes the following spring which will give the colonies a real boost. It is believed that by providing your heather-going colonies with combs of heather honey from the year before, the bees will go out foraging days ahead of the colonies not so provided for.

The late Brother Adam [4] was convinced that colonies taken to the moor should always be headed by queens in their second year and that the excitement of the move actually stimulated egg laying,

Most heather-going beekeepers use queen excluders on the moors, for starters and ekes will result in a lot of drone comb and queens are particularly attracted to laying in the larger drone cells, even late in the season, if they are available. This would result in many spoilt combs which should have held honey. There are occasions, however, particularly in late seasons, or when the nights are cold and the bees tend to move out of the supers altogether, when queen-excluders [5] can be dispensed with and removed. Experienced beekeepers often obtain some combs of heather honey this way when other beekeepers get nothing in the supers.

Traditionally, heather comb honey was produced in sections on the moors and the small racks holding no more than 18 sections fitted inside an empty brood chamber - used as a super - which was warmly packed with at least three folded hessian sacks. This helped to provide extra insulation thus keeping the hive warm [6] which is necessary for comb building and sealing: the bees had to be encouraged by every possible means. Section racks, holding up to 32 sections, which covered the entire brood chamber as a super, were, as a general rule, useless on the moors and many beekeepers were discouraged for good, when trying to work with them. A wide shallow frame carrying 3 sections was sometimes used and in recent times ROSS ROUND combs have been tried by some. But the simplicity of cut comb heather honey produced in shallow frames - Manley and Hoffman - outweighs them all. The narrower spacing of the latter is popular where families buy a few complete combs at a time and return the empty frames to the beekeeper the following spring.

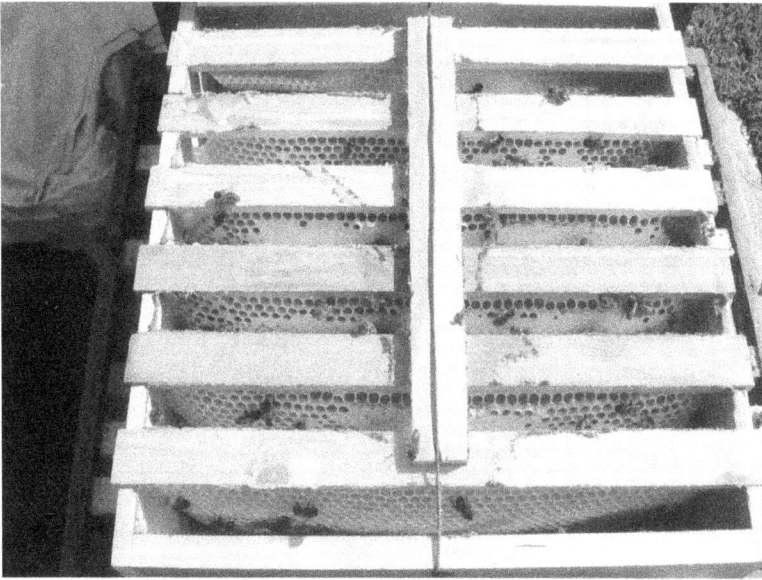

Heather honey produced in ekes is not as popular today as it was in former years.

Good, sealed combs of heather honey - some of my 2005 harvest.

Hives and their Preparation

National and Smith hives are ideal for heather work as the small brood chambers of both ensure that having them reduced from two brood chambers to a single brood chamber at the beginning of July you have an enormous force of foraging bees. As the brood emerges, the bees store honey in the brood combs and this brood chamber is removed altogether at the end of the month - with the aid of a clearer board if all the brood has hatched. If there is still sealed brood in some of the combs then the bees must be shaken or brushed off the combs with a goose wing feather into the bottom brood chamber and the combs without bees used to 'boost' nucleus or smaller colonies. The bottom brood chamber will be immensely crowded and a super of shallow combs with the frames fitted, if possible, with full sheets of medium thickness worker base foundation unwired, or, with starters of the same, or, drawn-out combs if you have them, are given to the colony above a queen excluder. The inner cover board is then replaced, ideally, with some form of packing or insulation immediately above. One super is enough to start with. The mistake so often made is to provide too much room at the heather. Two supers are often provided and invariably there are a number of unsealed combs in the second. If you are working for choice cut-comb heather honey it is much better to have as many combs as possible sealed. For those beekeepers 'geared up' with moisture extraction facilities, unsealed heather honey no longer presents a problem and they are quite happy to get as much heather honey as possible even if it is unsealed.

National and Smith hives are ideal for heather work. Here John Gleed is checking on the bees in his national hive which are working well in their second super.

The brood chambers of blossom honey removed before going to the heather are piled up on hive stands with clearer boards between each and with a hive roof on top: these can be returned to the bees for over-wintering. They should also be Good, sealed combs of heather honey - some of my 2005 harvest.When reducing double brood chamber colonies to one brood chamber, the problem of crowding all the bees into a smaller area must be tackled. Two supers can be used - the top one containing drawn comb - previously cleaned by the bees and dry. The other super immediately above the queen excluder and the brood nest contains shallow frames fitted with unwired medium worker brood foundation or starters of the same. These are topped with a screen board with a deep rim on the bottom part which will usually accommodate the large force of bees - when the bottom entrance to the hive is completely closed with an entrance block or strip of foam. Many experienced heather-going beekeepers leave the entrance open to allow the bees to cluster up the front of the hive, leaving the entrance block with its 12 inch aperture in place. The beekeeper must have considerable confidence to tackle moving bees in this way. Bee veil, and suit, rubber boots (Wellingtons) strong leather or industrial gloves, to prevent tearing hands on projecting metal parts, namely, hive tops, excluders, broken hive staples which have at various times been used to attach floorboards to brood chambers along with functional smokers; are all basic requirements. It is an advantage to have two strong hardwood lats screwed to the bottom of each floorboard to provide additional strength and to facilitate the lifting of hives.

The use of strapping is currently the most popular means of securing hive parts, but often beekeepers complement this by securing the hive floorboard to the brood chamber with hive staples at each corner. Hive entrances are best closed whilst doing this to prevent bees pouring out to sting the operator: the entrances can be opened immediately afterwards. Invariably, when loading hives manually and pushing them along a trailer floor, something will come apart. Carry something like Bluetack and small pieces of foam rubber to plug holes, etc, from which bees can escape. Buckfast Abbey carried toilet tissue for this purpose when moving bees to the heather. In heather districts some beekeepers attach hinged floorboards to their hives with ventilation holes covered with strong perforated zinc or wiremesh. The Scottish firm STEELE & BRODIE listed such a floorboard in their catalogue for years - with a thumbscrew permanently attached below to aid easy closing. Such hives should always be set on stands, rather than on the ground, as growing heather and bracken, can, on occasion, close the entrance altogether unless a small stone or piece of wood has been placed in position to prevent this.

A   B

A.

Colony on single small brood chamber holding 9 BS combs and dummy board with super of 8 Manley shallow frames at the heather. Note brood combs being filled with stores for winter.

B.

The same colony in mid-April of the following year, with combs almost all filled with brood, being provided with second brood chamber for the Oil Seed Rape with the frames fitted with wired worker base foundation.

D

 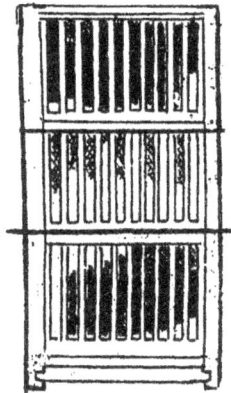

C

Early June when the second brood chamber - like the first - is filled with brood.

D

End of June. Second brood is raised above a super(s) of empty combs to the top of the hive, with a queen excluder between them and the bottom brood chamber

There is an equally effective system of colony management for the heather for those who prefer to run their bees in small, single brood chamber hives on 10 BS combs. This system is ideally suited to those beekeepers who work on their own, where the lifting of heavy hives can be a problem and the removal and manipulation of several brood chambers and the concentrating of a large foraging force of bees into a single brood chamber is off-putting.

In late April or early May when colonies have over-wintered well and are expanding beyond expectations on the nectar flow of the Oil Seed Rape, do not super. The congestion thus created will inevitably compel the bees to 'put up' a few choice queen cells. In districts where there is no nectar flow from the Oil Seed Rape these conditions of prosperity can be brought about by the heavy feeding of sugar syrup. When these conditions are met, a good five-comb nucleus colony can be made up in the empty brood chamber of a similar type of hive, alongside the other. Ensure that the old queen is left in the parent colony and add drawn comb, if available, or else frames fitted with foundation, to the flanks of the remaining five combs.

The nucleus should have two frames of food and three combs of emerging brood in the middle, densely covered with bees and with the queen cells too. On each side of the nucleus put a dummy board. Take care not to shake or jar the combs with queen cells , and ensure that the hive has a small entrance, which must in any case be completely closed with tightly packed grass for several days which, as it dries out, the bees will remove. No bees must escape to return to the parent colony alongside until the bees in the nucleus make their own way out. Once the bees achieve this and begin working normally from the small entrance, the surviving young queen should soon mate and a check 14 days later should establish this. By the end of May it should be possible to boost the nucleus colony with a comb of emerging brood from the parent colony alongside and continuing at fortnightly intervals throughtout June until the full complement of 10 BS is made up. By mid-July the nucleus colony - with the young queen - should be packed with hatching bees and in an ideal condition to be taken to the heather. At this stage it is provided with a super. On a warm afternoon - when there is a nectar flow in progress - move the parent colony away on to another stand in the apiary. The returning foragers will further boost the nucleus colony built up during the summer, Really 'topping' heather-going colonies are obtained this way. There can be uniting with newspaper upon return from the moors if the object is to keep the number of honey producing colonies the same in the apiary, whilst maintaining a high level of production.

Using Swarms for Heather Colonies

Traditional heather-going beekeeping in many parts of the country is now intrinsically tied up with coping with winter sown Oil Seed Rape which flowers from April through to June, and then the spring sown varieties which flower in mid-June and July, a small brood chamber of 10 B.S. combs, (15) use of a queen excluder and the lack of timely supering. This results in a spate of swarming from mid-April onwards. Often, swarms 'come off' without queen cells being started in the hive. A new generation of beekeeper has appeared to take advantage of the situation and they collect all the swarms they can and set them up in hives and, if prudent, treat them for mites. These should be hived initially on 5 British Standard frames fitted with sheets of wired worker base foundation with dummy boards on the outsides of the frames and some loose sacking in the outside space to prevent the bees from building wild comb. A feeder should be added, and some simple packing placed above the bees to conserve heat. New, additional frames are added as the original complement is drawn out. By the end of July these are usually 'topping' stocks to go to the moor on a single brood chamber and one super.

Moving Bees to the Heather

The move to the moors should always be made at break of day - loading up to be away by 5 am - reaching the moors before the powerful rays of the sun can be felt - as many powerful colonies of bees are lost over the years from overheating and suffocation [7] There is no more distressing sight than to lose a top colony of bees this way - which usually involves the collapse of the combs and honey pouring from the hive entrance. Unless it is a cool damp evening and you know exactly where you are going to, and you are not going to get stuck, moving bees at night can be fraught with difficulties - as during the warm evenings of late July and early August - the bees are often still returning to their hives as darkness falls and many actually stay out all night on flowers and return the following morning when the hives have gone - much to their distress. If you have some colonies still left at home, however, these 'lost' foragers will usually join them.

John Phipps moves his bees very early in the morning to be on the site by dawn. Straps were used to hold the hives together during transit and the hive entrances blocked with foam rubber.

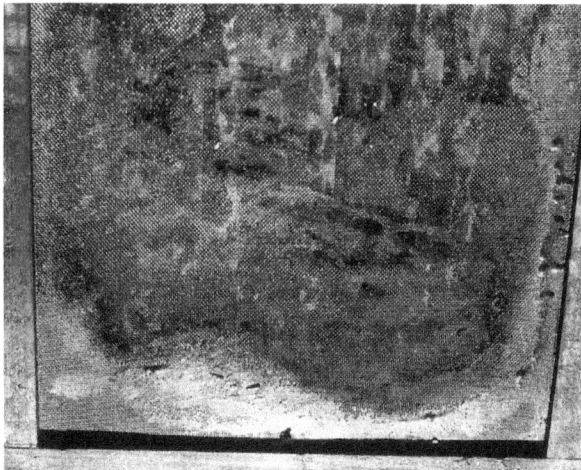

Wire screens were placed on all the hives, but each year they become covered with propolis. It is essential that the screens are cleaned before putting them on the hives. Once on site the screens are often left in place and a crownboard fitted above them.

Grass, such as lawn mowings, along with damp moss, is excellent for closing hive entrances if the move to and from the moor is only a short one, as it has the additional advantage of allowing the bees to eat their way out of the hive should the beekeeper overlook this. Unfortunately this happens somewhere in the country, every year. Having arrived on the moor safely with the bees and got the hives onto their stands a feeling of satisfaction prevails and the possibility of a record crop of heather honey is within reach. Lulled into a false sense of well being, the beekeeper and helpers sometimes set out for home without double checking each hive to ensure that the bees in every one can get out. It is worth returning to the heather stance later in the day to ensure that all is well.

The bees are released by withdrawing all the foam strips. The strips are counted to make sure that they are the same number as the hives. Walking up and down a row to check on the hives from which the bees have been released can be quite exciting but not recommended - at least until the bees have had chance to settle down.

The Use of Other Hives and the Problem of Unsealed Honey

The small single brood chambers of National and Smith hives are ideal for heather work [8], ensuring that much of the honey goes into the single super and is sealed. By the end of August the bees store honey in the brood combs - as the brood emerges - and bee colonies taken to the heather moors are usually well-provisioned for the winter. As experienced beekeepers know, the high protein content [9] of this honey ensures that these colonies will have more 'zip' to them than the colonies that have stayed at home and been over-wintered on blossom honey - a large part of which is often crystallised Oil Seed Rape honey. The National hive - improved by the late Arthur Abbot of Mountain Grey Apiaries, Brough, Yorkshire, with its splendid brood chamber and super handling features which allow them to be carried easily by beekeepers on their own - a great improvement on the ' cups' scooped out of the sides of hives assembled from four pieces of wood. Those beekeepers which have such hives screw a length of wood on each side of the brood chamber and super to facilitate lifting. The Smith hive [10] is often described as a small Langstroth. Beekeepers, such as Athole Kirkwood, have had success with the standard Langstroth hive on the moors whilst on the North Yorkshire moors John Whent operates a thousand or more colonies of bees working with standard Langstroth equipment and using brood chambers as supers. With such a large comb area there is often much unsealed heather honey - but, with modern moisture-extracting equipment the moisture content of the honey can be reduced down to an acceptable 20% and unsealed heather honey is now no longer the problem it once was. Many small-scale beekeepers use simple moisture extraction devices in their honey stores. Spare bedrooms, too, with central heating, are sometimes used to store heather honey in the supers [11]. The Commercial hives, with their 16" x 10" frames, are used by other beekeepers with some measure of success at the heather, whilst others work with the even larger Dadant. These larger hives are extremely heavy when filled with heather honey, and the crops from simpler smaller hives can be equal and often better at the heather.

Over-wintering Colonies

For successful over-wintering, the honey from the Ling must be sealed and the bee colonies sited so the hives catch the mid-day sun, to encourage the bees to take cleansing flights in mid-winter. A top entrance to the hive can be an advantage as well. All too often there can be much unsealed heather honey in the combs and, unless such colonies are fed heavily with thick warm sugar syrup, there will be problems galore from fermentation. Honey from the Ling is notorious for absorbing moisture and, if left in the unsealed state on the hives, it will soon be frothing in the combs which is lethal to the bees. I have on a number of occasions visited moorland apiaries at the request of beekeepers where heather honey, much of it unsealed, had been left on the hives, to find brown excreta running from hive entrances. This unpleasant smell will always stay with you. In severe winters when the bees have been confined to their hives for weeks on end, the first cleansing flight will be remembered for the pungent smell of heather honey in the apiary, as the bees on the wing void their faeces.

When supers have been removed and the hefting of the hives with the roofs off reveals that they feel light in weight and have few reserves of stores in the brood chambers, then such colonies must be provided with slabs of fondant placed immediately above the bees. Checks must be made throughout the winter to replenish the food if necessary.

The Best Bees

Strains of the north European dark bee (*Apis mellifera mellifera*), adapted to the district, give over a period of years the best performance at the heather [13]. Many of the dark bees found in the British Isles stem from the extensive imports of Dutch bees, Ligurians from Northern Italy, and French Blacks which the Scottish firm of Steele & Brodie made available to British beekeepers from their supplier, William Wilson. These LE GATINAIS Bees from Grigneville, Loiret, and Faronville, France were, apart from their uncertain temperament, highly thought of amongst comb honey producers in this country. An advert from the firm in 1927 stated "As recommended by J M Ellis of Gretna Green whose two stocks produced 240 finished sections which sold for £20" Both the late Brother Adam and Ted Humphreys recalled that for heather honey production these bees had no equal. But, on a number of occasions, these gentlemen had to retreat from the bees' fiery onslaughts. Brother Adam found himself standing half-submerged in the mill race which meanders through the Abbey grounds in order to escape from the bees. Thornes, too, had their own supplier. Like many other beekeepers I have tried light-coloured American and New Zealand bees over a period of 50 years. The extreme docility of these bees is their main claim to fame, but for serious heather work they should be avoided. They will certainly fill supers of heather honey in their first season, but invariably the colonies collapse with Acarine infestation and Nosema during the winter months and following spring. The crosses from the colonies that do survive are, as a general rule, unpleasant bees to work with as Woodbury observed in the 1850's when he first introduced Ligurian bees into Devon [12]. Charlton, who started the Northumberland BKA in 1888, favoured the Ligurian bee for out-crossing as many colonies of the local dark bee had serious defects of the brood which the introduction of the Ligurian corrected.

William Herrod - Hempsall, Brother Adam, Manley and Gale [14] all enthused, at various times, about the virtures of the yellow bee when they were obliged to restock following the enormous losses to the honey bee population of this country up to, and following WWI. It is now thought that this was brought about by a combination of factors, including the mite Acarapis woodii along with viruses and poor beekeeping practice.

Following the adoption of the movable comb hive, which allowed beekeepers to remove almost all the honey from the bees and replace it with sugar syrup and candy, Manley gave beekeepers a clear warning of the folly of this practice in his book BEEKEEPING IN BRITAIN (1948) page 367. This, along with his remarks about the brood defects of bees, should be carefully considered today. William Herrod - Hempsall, after losing the dark bees which he so much loved, worked with Root's Red Clover strain from the USA and developed a strain of yellow-banded bee which, surprisingly, could be accommodated on 10 B.S. combs. However, with these bees he went on to encounter serious over-wintering problems which he attributed to Nosema.

Problems on the Moors
A. *Starvation due to Poor Weather*

There are seasons of complete failure on the moors such as 1912, 1946, 1954 and 1985 when many bee colonies were lost from starvation. I well recall going round the stances of many northern beekeepers during August 1946 and August 1954 and finding the ground in front of the hives carpeted with crawling, dying bees [16]. Milk churns filled with sugar syrup were taken to the moors, and wherever possible the crawling bees were swept up on to shovels and dumped into the tops of nearby hives then lightly sprayed with syrup. After an hour or so, when there was some response from the bees and they were gently humming, the old type of round feeders were put on every hive. The feeders were replenished every other day and eventually, from those colonies that had not died out, a small amount of heather honey was stored in the brood combs .With further heavy feeding of the colonies upon their return home, some of these stocks of bees managed to survive the dreadful winter which set in around January 20th, 1947, and continued until April of that year. However, honey again poured into the hives in August and September, 1947.

It would appear that hours of sunshine play an important part in the equation when the colonies are on the moors. Similarly, after the dreadful summer of 1954, the bees stored a small amount of heather honey in September of that year and then, virtually, no more honey was taken into the hives until July 1955 when an altogether fantastic nectar flow went on and on into September. Again, hours of sunshine played their part in this. Sometimes, during a heavy nectar flow at the heather, honey / nectar will be deposited in cells containing eggs, submerging them completely.

B. Heather Beetle

In some years the Heather Beetle, *Lochmoea suturalis*, becomes a nuisance[17]. Dr Guy Morison described the activities of the beetle in 1936 . A swarm of beetles had been seen to land on the waters of Loch Awe in Argyllshire and were speedily devoured by the trout as they struggled on the surface. Their habit is to take wing in swarms in Spring and travel wherever the wind takes them - which may be up to two miles and cover hundreds of acres. In Britain, and particularly in Scotland in 1935, severe damage by the heather beetle was widespread, the west and south-west areas appearing to suffer the most. Heather of any age is liable to attack, but whereas young vigorous heather up to eight or ten years is seldom killed, old heather of twenty years or more often succumbs to the attack of the beetle and on the moors where severe damage occurs most of the heather is very old. During the 1970's the heather moors of the northern Pennines between Blanchland and Stanhope (where, incidently, there are some fine stands of Bell heather, Erica cinerea) were attacked by the beetle on several occasions.

C. Shrimp Brood

One condition of the brood which I have seen on a few occasions during my 50 years of heather-going beekeeping was first described by the late Bob Couston as SHRIMP BROOD (page 72 "THE PRINCIPLES OF PRACTICAL BEEKEEPING"). During the months of June, in 1957 and 1962 and in early July 1975, there were heavy death rates amongst sealed brood in hundreds of apiaries throughout the East of Scotland. The worst colonies had over a third of the sealed brood affected, all of which had died on the 13th or 14th day stage of development, having the body segments formed and the pigment of the eyes just starting. The dead pupae were quite firm - almost crisp - and were reminiscent of the consistency and appearance of small, freshly-cooked peeled shrimps. The affected cells were scattered about in random fashion - some being next to normal pupae of the same age. In all cases, the malady cleared up within a three-week period of the onset of the condition but, on observing this in their apiaries, many beekeepers became worried because the initial appearance with dark and sometimes perforated cell-cappings resembled that of AFB. The outbreak always followed a poor heather season the year before - and it is interesting to record that not one case of the malady was seen amongst the thousands of stock inspections during the intervening years. There is certainly an opportunity here for Rothamsted or Sheffield to do some serious work on this.

Removal of the Crop

Early in September clearer boards are put on the hives during late afternoons. These, which hold several porter bee escapes, are added making sure that the springs are functioning and not stuck with propolis. An alternative is to use the Canadian cone escapes or the Hexagonal and Rhombus escapes listed in the bee appliance catalogues. Some beekeepers use Benzaldehyde on an absorbent pad - on a cool day - to drive bees from the supers. When using clearer boards the supers, hopefully cleared of bees, will be ready for collecting early the following morning. Set out at the break of day when it is still cool on the moors. If left too late, it can suddenly become hot, then the smell of heather honey will incite the bees into a frenzy and an unnecessary spate of robbing is sparked off. Load the pickup or van quickly and depart. There may be some colonies where the bees have not left the supers and it will be necessary to shake the bees off the combs. This is best left until another day when the bees have settled down and the clearer boards can be removed, the travelling screens put back on under the roofs and the hives made secure and ready for lifting. This is best done during daylight hours on some damp, cool day when all the bees are in their hives and the entrances can again be closed quickly with moss, strips of foam rubber, etc. Never use bee-blowers to get bees out of the supers on the moors as the stress caused to the bees invariably brings about over-wintering problems associated with Nosema.

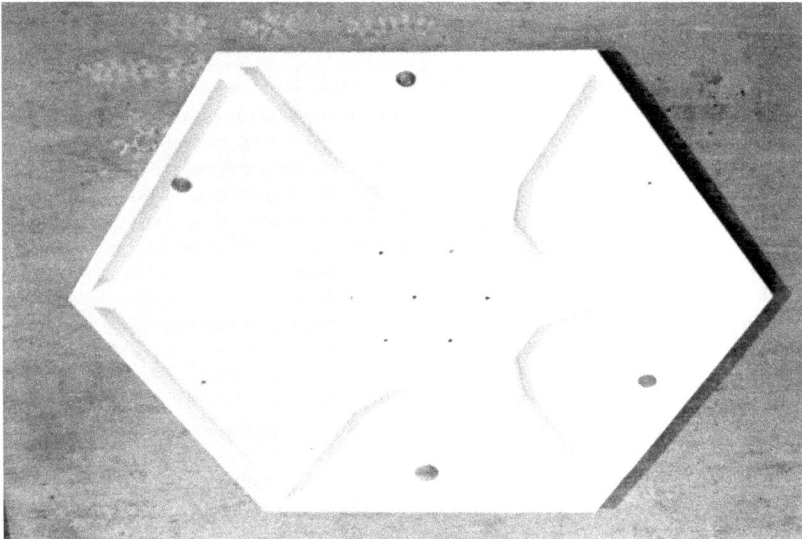

An hexagonal bee escape with many exits and no moving parts.

Frames of heather honey. If there are many unsealed cells then the combs may need to be placed in a warm room for the honey to 'ripen' as a high moisture content will lead to fermentation. (Photo: Job Pichon, Brittany).

When the hives are small, with only one super on each, the usual practice is to leave the honey on the hives and bring them off the moors as they are. Leave the honey until the first sharp night frost, for then the bees will have left the supers of their own accord as they go into cluster. The supers can then be lifted off the hives the following morning, clear of bees, after first breaking the seal between the brood chamber and super with the hive tool. Fastening devices are best removed as soon as the hives are brought back to their permanent apiary sites. Should straps be left on hives during the winter, shrinkage can damage hive parts. Entrance blocks with 12 inch wide and 3/8th high apertures will usually deter mice.

It is important to watch out that Braula larvae and Wax moth don't ruin the heather comb honey [19]. Honey from the Ling is extremely sensitive to overheating and freezing, so take great care when handling it. Pressing heather honey is a time-consuming business but the MG Press introduced by the late Arthur Abbot has invariably given years of sterling service since it appeared 50 years ago. In more recent times, STEELE & BRODIE made a stainless steel water operated press available. It is essential to have a really good mains pressure water supply for this. Larger producers of heather honey make use of mechanical full frame looseners which enable them to extract the honey tangentially. Smaller producers sometimes use a heather honey roller and spin drier for extraction in a room which is warm and dry and where, if possible, the supers of heather honey have been stored already for several days. Heather honey extracted in this way is certainly not as attractive in appearance as that pressed slowly in a MG, PEEBLES or STEEL & BRODIE Press.

A modern heather honey press produced by E H Thorne of Wragby, Lincolnshire.

Marketing Honey

Small-scale beekeepers usually dispose of their cut-comb and pressed honey to business colleagues who are anxious to obtain such a wonderful natural product, rich in protein, off the hills [20]. Larger producers of heather honey often market their crop through Co-operatives and Packers or aggressively sell their product at Country Fairs and Agricultural Shows. Northumberland's first commercial beekeeper, Robert Reed, who worked several hundred skep colonies around Acklington and Morpeth, between 1760 and 1812, was a pioneer in such marketing, attracting large crowds at fairs as, completely unprotected, he drove his quiet brown bees from their skeps and then scooped them up with his bare hands to fill empty skeps.

Rather than producing cut-comb honey or square sections, some beekeepers opt for using Ross Rounds. Once taken from the hive there is little handling involved in getting the product ready for market.

Control of Varroa

The small-scale beekeeper can obtain some measure of control by systematic removal of combs of sealed drone brood from colonies - during the active season - replacing them with drone base foundation. Other beekeepers will alternate the use of Bayvarol and Apistan strips in their colonies but only after the supers of surplus heather honey have been removed. Keep in close touch with MAFF for guidance on the clearance and release of other Medicants.

Labelling of Honey

The BBKA reminds us that indirect misdescription is illegal, eg, giving the name of a plant such as heather or showing an illustration of it, on a label when the container does not hold predominantly heather honey. The term 'predominantly' is not defined, but a proportion of at least 75% is thought by many to be reasonable. If in doubt consult your local Trading Officer.

Some past advice from STEELE & BRODIE: "Whether honey is eaten by family, given away, sold at the door, marketed through retail outlets or prepared for the show bench, it must always be clean and strained of all foreign bodies. Increasingly, as more chemicals are added to other foods, honey must be able to stand the scrutiny of Public Health Officers and an evermore demanding, health conscious public."

St Ambrose - Patron Saint of Beekeepers - A carving from Eastern Europe

NOTES - AND REFERENCES & PAPERS RELATING TO HEATHER HONEY - in the Scottish Beekeepers Association - MOIR LIBRARY - housed in Fountainbridge Library, Dundee Street, Edinburgh:

1: A.E.McArthur: WELL-KNOWN BELL HEATHER AREAS IN SCOTLAND - Scottish Beekeeper -Editorial -July 1996.
James Cunningham W.W.SMITH - Scottish Beekeeper - June 1969.
Dr. C.L. Whittles: Survey of Heather Areas - Scottish Beekeeper - September 1950.
Anna Maurizio: THE HEATHER HONEYS OF EUROPE 1973 SF 539 18.
J.Louveaux: HEATHER - 1973.
A.S.C.Deans: SCOTTISH HEATHER BEEKEEPING (St.Andrews) 1972.

2: HEATHER HONEY, Major Sitwell's lecture to the BBKA 1912: This made the British beekeeper take an interest in this honey. ISBN 0 95168 7 5. Sitwell first to make use of the word 'stance' to describe a heather apiary.

William Herrod-Hempsall's BEEKEEPING NEW & OLD, Chapter 12: Vol 1. An interesting account of 'old time' heather-going beekeeping. Whilst having no personal heather-going experience himself, Chapter 12 was the Author's own favourite in this great two-volume work on beekeeping.

3: Ted Humphreys: CLOUDBURST ON THE HEATHER MOORS - Bee Craft - February 1951.

4: THE PRODUCTION OF HEATHER HONEY, Brother Adam's lecture to the Newcastle & District BKA: December 1934. Had great success with a single 10 B.S. frame brood chamber, which had been reduced from 20 combs, on Dartmoor. Favoured French Black bees along with Heath bees from Germany and Holland. Printed privately 1934. Re-issued 1990.

5: Ellrig: NO QUEEN EXCLUDERS AT THE HEATHER - Scottish Beekeeper - November 1978.

6: William Hamilton: THE ART OF BEEKEEPING - 1945 - pages 121 - 128 . Hamilton makes the point that a small brood chamber of 10 B.S. combs, reduced from 20 B.S. combs or more, is essential for success at the heather - with as many bees of foraging age as possible crowded into the hives. Adequate packing required to conserve heat if completely sealed combs of heather honey are being sought. Favoured French and Dutch bees, or adapted local strains, for moorland work.

7: Rambler: MOVING BEE COLONIES - Scottish Beekeeper - April 1980, 1985 & 1986.

8: THE B.S. FRAME HARD TO BEAT AT THE HEATHER - S.Beekeeper - October 1968.
Willie Robson: WEATHER FORECASTING WITH BEES - Scottish Beekeeper November 1978.
Dr.John Anderson: FIRST PERSON TO HOLD HONEY JUDGES CERTIFICATE SBA 1919.
A.E.McArthur: Reports that Dr.Frederick Ruttner introduced the VARROA mite into Germany & Europe in 1972 - Scottish Beekeeper - September 1979- Subsequently, first reported in England, at the Saffrey's apiary at Cockington, Torbay, Devon: on Saturday April 4th,1992 at 2.30 pm.

9: J.Pryce-Jones: HEATHER HONEY - Bee World - 1936 & 1950- Scottish Beekeeper 1941.
Solveig Lund: DRONES TRAVELLING 40 MILES & SPREADING VARROA MITES -Scottish Beekeeper - July 1998.

10: John Ashton: A CONVERSATION WITH W.W.SMITH OF INNERLEITHEN. Interviewed by Selby Robson. Published by Beekeeping Centre Kirkley Hall, Northumberland: March 1962.
C.Neil Anderson: W.W.SMITH OF INNERLEITHEN - Scottish Beekeeper - May 1980.
SEASONS OF DISASTROUS LOSSES - September 1987.

12: George Charlton: BEEKEEPING IN NORTHUMBERLAND - Hexham Herald.

13: H.E.Hitchen: The Leics beekeeper who contributed to the BBJ and Bee Craft in the 1930's believed that honey bees adapt to the district they work in and were greatly influenced by environmental factors: and that it was essential to develop local strains of bees: BBJ - April 1934.

14: R.O.B. Manley deals with heather honey in HONEY PRODUCTION 1936 & 1947 - Chapter 17. HONEY FARMING - 1946 Chapter 10. BEEKEEPING IN BRITAIN 1948, pages 305-12.

15: Donald Sims: SIXTY YEARS WITH BEES - 1997 - Chapter 11 - Heather Honey ISBN 0-907908-799.

16: HEATHER FLOW IN SEPTEMBER - Scottish Beekeeper October 1986: Athole Kirkwood: 1985 THE WORST SEASON IN 40 YEARS - Scottish Beekeeper - Jan 1986.

17: THE NORTHERN APIARIST: Journal of the Northern Federation BKA's 1945 - 1954. 887 - Re-issued, 1990.

19: George Green: HEATHER HONEY - Scottish Beekeeper - July 1955.

20: Captain Thake: CUT COMB HEATHER HONEY-& GRANULATION OF HEATHER HONEY AFTER PRESSING - Scottish Beekeeper: November 1971 - page 190.

Neil Price: DEMONSTRATES HIS COMB CUTTER - Scottish Beekeeper: August 1972.

Bob Couston: MICROWAVE TREATMENT OF HONEY -INCLUDING HEATHER -.Scottish Beekeeper April 1982.

21: I.A.Khalifman - Pchelovodstvo: No 8 - HEATHER BEEKEEPING IN NORTHUMBERLAND 1958.

John Phillips THE CASE FOR HEATHER RASE Journal - 1990 - pages 96-102.

Cecil Pawson: RASE SURVEY OF NORTHUMBERLAND - 1961 - Chapter 10 BEES pages 143/4.

Joseph Tinsley: BEEKEEPING -1945 Chapter 11 - Production of Heather Honey.

Chapter 12
A Selection of Edited Extracts
from Colin Weightman's Columns in the
British Bee Journal

One hive four queens
(BBJ 1989)

I once had an old queen and three daughters all laying together in the same colony! The three queens went into winter, but by the following May only one of the young queens was left. I reported this in a letter to the Scottish Bee Keeper in 1946.

Winter of 1962/1963
(BBJ March 1963)

Snow covered hives at Shilford, 1963.

The present winter now exceeds the much-talked about storm of 1947 both in duration and severity.

At last, during the late afternoon of Monday, March 4th, warm air currents from the Atlantic pushed their way across the Pennines and forced back the high pressure which had maintained wintry conditions without a break since

December 28th. And by the simple process or rising temperatures, along with wind and rain, the lowlands of Northumberland were almost cleared of snow by March 9th: a total of 74 days (with November, 1962, 80 days). It was a dramatic thaw, but in many places it brought disaster.

Now that the days are mild we have made a round of the bees and have reduced hive entrances from full width and changed the floor boards. Despite the severity of the winter we have had 100% wintering. It would appear that providing stocks are favourably sighted to catch all the winter sunsnine there is, and are provided with the right kind of shelter and protection, and the colonies themselves are normal in every way and left with ample stores, bees can stand up to any of the rigours of the British climate.

An interesting feature this spring is that colonies, without exception, are occupying top chambers only. Invariably, at this time of year, with double brood chamber lots, we find the clusters situated between the chambers, and for the first time in years, no drones have been carried through the winter by these powerful lots of bees.

Despite ghastly conditions the lambs are managing to survive. In Northumberland, our own ewes close to lambing have been brought indoors and the lambs dropped are frolicsome and making headway.

At Buckfast, the Rev. Brother Adam says, "The weather has, indeed, been exceedingly wintry here since Christmas - with as much as 24 degrees ot frost. On a number of days we had eight degrees of frost at noon. I have no idea how the nucleii are faring at the queen-mating station. Owing to the ice and the deep drifts of snow it has been quite impossible to get anywhere near the site."

My own nucs high up on the Northumberland moors were, according to two shepherds, covered by deep drifts of snow for weeks, so feared the worse. In the valley apiaries bees have been allowed occasional flights and have cleared the dead, etc, and the clusters have readjusted themselves on stores and fondant. There has been considerable interest in the variations of fondants available. We have had the best results with fondants made from white sugar, cane sugar, with a low water content.

The present winter now exceeds the much-talked about storm of 1947 both in duration and severity.

The New Year of 2011, once again brought heavy snowfalls to much of Britain including Colin's apiary at Shilford.

Winter Wonderland
(BBJ February 1981)

January 1st this year was a jewel of a day. I crossed the moorland tops which separate the counties of Northumberland and Durham from Cumbria and Yorkshire to see the condition of hive stands at the heather stances, as repairs have to be made every year or two before the bees are moved up to the moors in August. This stark lonely region was a winter wonderland. Crisp snow covered everything and in the distance Mickle Fell - Yorkshire's highest point - stood out in the winter sunshine. A stoat in winter ermine crossed the road and as I returned home I thought of the late Arthur Ransome's lines in Winter Holiday:

"Softly at first, as if it hardly meant it, the snow began to fall."

An Early Spring?
(BBJ February 1983)

For the first time since 1949 Snowdrops have come into flower here during the second week of January and the white drifts remind me of Nell Darby's descriptive lines "Like tiny nuns in green and white". In contrast to last year the winter-sown crops are a brilliant green. Moles are working overtime all over the place and the various rabbit catchers tell how the now abundant does are all carrying young.

During December we were reminded that the local fox population was alive and well as the penetrating cries of a vixen could be heard nightly trying to attract a dog fox. These activities usually occur here towards the end of January and our farm collie and terriers responded fully to this moonlight serenading.

Wet Summer of 1985 Responsible for Winter Losses

Following the wet summer of 1985 and the loss of bee colonies in the spring of 1986 it was reported that bee colonies in Northumberland were reduced from 3,000 to about 1,500. Numerous queens were brought into the country from New Zealand - as supplies from USA came to an end with the appearance of the Varroa mite there. By 1987, the "ginger" coloured crosses were being noticed in apiaries which had had dark bees for years - along with the marked susceptibility of the New Zealand bees and their crosses to the tracheal mite.

Swollen Streams Flood Apiary
(BBJ December, 1980)

During days of heavy rain, in one afternoon one and a half inches fell in twenty minutes. Water cascaded off the hillsides and the moorland streams soon overflowed. Ted Humphreys, a Lancashire honey farmer with 160 hives, told me that the water in the Wyre, at Laangden, was 15 feet high and 50 of his stocks were swept away and the rest were well submerged. When the water had subsided, later in the day, Ted and his sons found sand blocking the entrances of the hives and the bees caught outside had returned and clustered on the outsides of the hives. In the following days, Ted and his helpers scooped these bees up and when placed on combs of stores many of the bees recovered. When the stocks were eventually examined, it appears that the bees had clustered tightly over the brood and had kept the water out. Two days later, combs of brood were found scattered on the river banks a mile and a half downstream and when salvaged some of the sealed brood was still alive and bees were emerging from the cells.

Freak Storm
(BBJ September, 1980)

The village of Fourstones - where the Hexham Honey Show is now held - was devastated by a freak hail storm at noon on Friday August 1st. After a hot humid morning which brought the hay makers out into the fields, storm clouds gathered and a sheet of ice particles came down from the heavens, stripping leaves of the trees, flattening standing corn crops and pulverising the haulms of potatoes in the gardens. I recall seeing similar devastation around Ripon when visiting the Great Yorkshire Show in July 1965.

The Heather Beetle
(BBJ November, 1981)

Owners of various heather moors in this area have expressed concern at the reappearance, this year, of the Heather Beetle. ThIrty five years ago, in 1946, areas of the local moors were devastated by this insect and it was widely held that the ling would disappear altogether as a result of the attack. But the much talked-about winter of 1946/47 followed and, for some reason, no more was heard or the beetle. The Northern Counties' own small bee paper, The Northern Apiarist, had just begun its nine year run and the then Editor, John McNichol, the Sunderland Optician, gave coverage to the heather beetle (*Lochniaea suturalis*). Writing in October 1945, he had this to say:-

"On my second visit (to the moors) I searched around, on the foliage and under the plants, until I found, literally, hundreds of small grubs. The foliage was nipped and part of the wooden stems bare, so I brought home some samples and sent them to King's College, Newcastle, and also to the North of Scotland College of Agriculture, Aberdeen. A few days later I received a letter from Dr R. A Harper Gray, Advisory Entomologist, King's College, that the heather example was no doubt attacked by heather beetle. Harper Gray and his assistant visited the area and reported that they found an adult beetle and a quantity of larvae. The pupae were in their state of rest, so there was no more harm experienced this year. Dr Guy Morrlson reported from Aberdeen that the sample of heather (*Calluna vulgaris*) received had been attacked by the larvae of the heather beetle.

Heather of any age is liable to attack, but rarely does the attack result in permanent damage. Beetles were found to be present on all kinds of heather moorland, but were much more abundant on wet boggy ground".

Drones

During the hot summers or 1988, 89 and 90 I carried out experiments on the Northern Moors which satisfied me that during the hot weather of July, drones were capable of flying up to three miles across the windswept moorland tops to find nuclei with virgin queens. This was established when 4 BS comb nucleus colonies were brought back to the home apiary twelve miles away. Returning to the moorland mating apiary each afternoon between 2-4 pm - when it was still and warm - a few drones from the apiaries of other beekeepers three miles away reached this secluded spot and continued to do so for a week until the weather broke - with rain.

Oil Seed Rape Honey
(BBJ May, 1981)

Ingrid Williams, in her informative article on Oilseed Rape (Bee World, No. 4, 1980, p143) tells us that "increased areas of rape should result in increased honey production." It appears that many beekeepers are climbing on the bandwagon, as they learn to handle the honey, and are moving their stocks to take advantage of this flow. Blended with other honeys it now appears to be accepted by the public.

Rhododendrons and Bees
(BBJ, 5th April, 1960)

A few weeks ago I referred to Northumberland's own expanse of rhododendrons at Cragside, Rothbury, in the picturesque valley of the Coquet. Just after World War II the beekeepers of the area had a problem: they claimed heavy loss of bee life when the shrub was in flower and that there were definite indications of poisoning. One beekeeper went further and said that occasionally he got comb honey in sections which when eaten caused dizziness and distress. I mentioned this at the Middlesex Summer School of 1958 and my statement interested Dr Alan Birch who later wrote to me, but unfortunately we were unable to corroborate it with the experience of the present generation of beekeepers at Rothbury. My colleague Harry Whitfield (who is well known on the show bench) whose district it is, definitely rules it out.

In June 1958, my friend Mr Morris (from Birmingham) happened to be staying with us and we made a special 'on the spot' investigation at Cragside. It was a wonderful sunny day after a week of heavy rain, the hillsides were aglow with colour and we drove the car slowly round the splendid estate. But we failed completely in our search to find honeybees interested in rhododendrons, although the bee colonies were particularly active.

The lesson to be learned from this is that in the past, under particular conditions, bees have worked rhododendrons on the estate, but now they seem to leave them severely alone. It will be interesting to see if they ever go back to them in this NE part of England.

Thefts of Queens and Colonies
(BBJ, December, 1980)

The nationwide publicity given to the losses of breeder queens, etc., sustained by our old friend Brother Adam this year, emphasises the trend among some people today to helping themselves to other people's property, As recently as 1978 Brother Adam told me that he was perhaps fortunate that in all the years the Abbey apiaries have been in existence they had not been vandalised.

For many years I maintained a queen mating apiary in a remote valley on the moors on the Durham/Northumberland border until the loss of nucleus colonies by bee-keepers helping themselves to the small colonies - they transferred them into travelling boxes - went well beyond tolerable limits. Only this week, my milkman sought me out to tell me that, he too, was a bee-keeper and that he had maintained an apiary at the Passionist Monastery at Ministeracres for a number of years until someone helped themselves to 17 stocks. Two stocks were left as the hive floorboards had dropped off and the individuals concerned had been obliged to put the hives down.

(In the 1950s the late Peggy Logan of Muir of Ord and I used to visit the Monastery to help the brothers with their bees and on one occasion Peggy introduced a Highland queen into one of the colonies using her favourite matchbox method.)

Crown Boards, Floorboards and Over-wintering Bees

Beekeepers who visit me are often surprised to find that I don't use wooden inner covers on my hives - with my potato-growing interests I have had access to fine hessian used potato sacks from the north of Scotland. I am satisfied that in my apiaries stocks with three folded sacks held in an empty super directly above the bees have always been in a more forward condition than those colonies which have had only an inner cover board immediately above the bees - even if such colonies have been provided with packing above the board. Peggy Logan, of Muir of Ord, had such thoughts too, and on her visits to me in the fifties she described how her colonies were housed with porous packing immediately above the bees. I get two seasons out of the sacks which are next to the bees and when there are signs of deterioration they provide excellent and fragrant smoker fuel, as that popular beekeeper Donald Sims of Foxton, Cambridge, will confirm, if the material is cut and rolled to the required lengths.

Floorboards - and the size of hive entrances - are all important, too. Floorboards generally available are too flimsily constructed and have a short life. I now attach two hardwood runners to the underneath side of each floor which at times of loading allows operators to lift the hives from the stands with ease. An additional advantage, hives slide well if you wet the bottom of the truck or trailer before loading. Entrance blocks have a nine inch width, 3/8 depth, entrance cut into them - which appears to deter mice - and provided the floor itself does not protrude in any way they last much longer than those floorboards which do project to provide a flight board.

Years ago, William Herrod-Hempsall, on his visits here, went to great pains to impress upon me the importance of having properly constructed floorboards and he favoured attaching a piece of cardboard or similar material on the underneath side of each floor and painting it. He describes this in the second volume of the book which was his pride and joy: Beekeeping Old and New (Page 1081). For a long time I doubted the wisdom of all this until some of the poorly constructed National and Smith floorboards fell to pieces when we were moving stocks to the heather. As there were no spare floorboards available - screen boards - which already had small entrances cut into the frame on the top side, and which were on hand, were brought into use and they served as floorboards on the moors. When we brought the bees back from Ihe heather we found that, apart from the entrances, these temporary floorboards had been covered with propolis and the holes in the screens immediately below where the bees were clustering were completely sealed.

We replaced some of the screen boards with wood floors when the hives were returned to their stands in the apiaries but six of the screen boards were left in place and these colonies were in very poor shape the following spring, the result of colonies being too exposed to the elements. The stocks provided with a wood floor overwintered well. Over the years I have had most success in the over-wintering of bees by providing colonies with reasonably small entrances in the bottom floorboards along with a top entrance - a small hole in the hand hold of the top brood-chamber - which can be closed with a cork when not required. Sometimes the bottom entrance has been closed completely. If the colony is occupying a single brood chamber - open bottom entrance - an empty super between the floorboard and the clustered bees is a great help too.

Shading hive entrances from the sun to prevent bees flying out across the snow again has served no useful purpose. The bees that go down on the snow are probably best out of the colony altogether. Attention to these particular points together with the other basics - ample stores and pollen; young queens; and a low incidence of adult bee diseases should ensure that your colonies are in an altogether stronger condition when the spring days do arrive. The days of finding miserable handfuls of bees awash in dysentery-stained hives, as many beekeepers were left with in the spring of 1979 - and after every severe winter - should now be behind you.

Illingworth's Oilcloth Excluders
(BBJ 1st February, 1967)

The present winter reminds me very much of the open winter 1948/9 which was followed by what some regard as the summer of the century. Another first-class season would do the Craft a power of good. In January 1949, I was staying in Cambridgeshire; it was then so mild that it was possible go outside without a jacket. It was during this stay that I called on the late Leonard Illingworth and his sister Mabel, at their home, "The Way's End", Foxton, near the Herts, border. Annie Betts was then still Editor of Bee World, and Leonard Illingworth Secretary of the Apis Club and BDI. "The Way's End" was a thatched property and the Illingworths kept geese, a large number of cats, and some 60 colonies of bees in Langstroth hives which they used to operate on a brood chamber and a shallow super system of management: the shallow super most seasons providing a satisfactory reserve of food. At that time though the Illingworth's main honey crop was from sainfoin. During the active season squares of oilcloth took the place of queen excluders- the bees being allowed access to the honey supers at the sides : the oilcloth was deliberately cut for the purpose as Illingworth had found that homecoming bees, coming from the fields heavily laden, ran up the inside walls of the hive. The queen, on her part, rarely found her way past the sides of the oilcloth.

Moving Hives to the Moors
(BBJ September, 1982)

When moving bees to the moors on a couple of occasions I noticed some interesting happenings. Because of humid conditions many of the older foragers 'slept' out all night and returned to the apiary around 9.00 am when the sun broke through the mist. I set travelling boxes out with a drawn comb in each to collect these stragglers and, later in the day, they were shaken at the entrances of several small 9 and 10 BS comb colonies with young queens, which were in the home apiary. It was interesting to see how these old foragers - many with pollen still on their legs - were accepted at once by the queen right colonies.

Not a bee was lost in fighting and the foraging force of these colonies was given a substantial boost. Supers of empty drawn comb were given the following day to accommodate the bees and these were subsequently filled with honey. I see that Bee World, vol 63, No 3, page 117, refers to the same thing happening in the USA.

MAKING UP NUCLEUS COLONIES
IN NORTHUMBERLAND

My bees are pleasant bees to work with and reminded Vlad Dermatopolous of the docile Cecropia lines in Greece ten years ago. I mention this as the late dear Beowulf Cooper, when he visited me in the early sixties, was anxious to hold and make full use of this important characteristic. My own "Robson and Cessford" line of local bee, which I have had now for 50 years, has retained its docility too.

But how far have we come in this direction? Visiting apiaries of some who have entered the craft during the past 25 years, the behaviour of the bees is quite unacceptable and one wonders how much this is tied up with over-manipulation?

The days of slow deliberate movements in the handling of bee colonies, avoiding crushing bees and jarring combs seem to have gone for good. Many of the beekeepers of today appear to be going into battle as the bees pour out of the hives as they approach them. With proper handling of the bees it is still possible to manage apiaries with only a protective veil and the correct use of the smoker.

To maintain this uniformity and temperament in my Robson and Cessford line, I made use of mInature Swiss mating boxes in the early years. These could be placed in woods and in isolated spots on the moors. However, setting up and maintaining these small boxes in our variable climate was too complicated as many others have found.

In 1948 I established a permanent queen mating apiary on the moors where the colonies were over-wintered on 7 and 8 BS combs This was finally disbanded in 1968 when the location became too well known and beekeepers helped themselves to queens!

From 1968 until the present time I have had most success by keeping a whole apiary of these bees entirely on their own. When colonies reach a prosperous state, usually towards the end of May, a nucleus colony of 6 BS combs is made up in a second brood chamber with division boards and placed above a framed screen board of perforated zinc with a small entrance, which in turn is placed above the parent colony and its supers with the small entrance at the back closed with a temporary wisp of grass.

When making up each nucleus, the queen in the parent colony must be seen and left in the bottom brood chamber. Four combs of hatching brood are taken to make up each nucleus along with two combs of honey and the bees from some of the other combs are shaken in as well. The combs from the bottom brood chamber are usually replaced with drawn out comb and it is essential to leave combs of honey on the outsides of the brood nest. A hatching queen cell is given to each 'nuc' from the breeder cell-producing colony selected for this purpose each year. Another selected colony is encouraged to provide a large number of drones by giving it several shallow combs of drone cells in March.

The advantages of this system are considerable. The nucleus colony benefits enormously from warmth from the colony below, and in our cold sunless summers there is no chilling of the brood and chalk brood is retarded,

With the traditional methods of making up separate nucs in the home apiary there is often a return of bees to the parent colony and subsequent robbing from this source. The chilling of brood too can be considerable. Best results are obtained when the newly made up nucs can be moved in their small hives to another apiary entirely on their own.

Using the framed screen boards the young queens mate well and quickly from the back or sides of the hive. There is usually no attempt at swarming by the parent colony. In the middle of July, the older queen can he removed along with the screen board and the colony reunited by the newspaper method (see Snelgrove's The Introduction of Queen Bees). After a week the combs in the colony can be reassembled to the particular requirements of the beekeeper. If you are young and strong and have helpers you can move the bees to the heather on two brood chambers - in addition you can get good combs drawn out in the top box. Others prefer to reduce the colony to a single brood chamber and crowd the bees. Whichever way one plays it, there will be a grand colony of bees which will provide one with a super of heather honey - even in seasons such as 1988. This is how I do it here and it works very well.

Country Life
(BBJ February 1983)

On the moors with Brother Adam

One sparkling day I visited the moorland farms where we take the bees at heather time. Brother Adam accompanied me on such a round in November 1981. It Is interesting to find that some of the farms have been modernised beyond all recognition while others remain unchanged and it is still possible to find, in 1983, the old farmhouse kitchen where the hens still have access and actually lay aggs in the house. At one such farm an overgrown pet lamb sleeps blissfully on the sofa and jumps down to greet visitors.

However, it was the sight of fine Rhode Island Red hens in the moorland farmhouse that reminded me of a local tinker who bought some choice combs of heather honey from Robson and Cessford of Riding Mill in 1951. The combs were stored, in their supers, in the living room until the very wet year of 1954. Even then the combs were still pleasing in appearance but the honey had set like toffee. Amid all the confusion of the living room, at least half a dozen hens sat together on the top rail of the fire guard while another hen occupied the empty fireside water pot which contained some hay and informed us, in no unceriatn fashion, when an egg was laid!

Mountain Grey Apiaries
(BBJ May, 1981)

Arthur Abbott or Mountain Grey Apiaries, Brough, Humberside, was destined to leave his mark on beekeeping in the British Isles. The early price lists of his firm show how he adapted the flimsily constructed National hives then available to the now immensely popular Modified National Hive. The MG Wax Extractor and the MG Heather Honey Press were other practical offerings to beekeepers. But my own favourite from the MG stable (apart from the Modified National hive) was the MG Parallel Radial Honey Extractor which I got from Arthur over 30 years ago. It has been in regular use ever since and Is still sound as a bell.

I last met Arthur Abbott in the early seventies when judging at the Hull Honey Show and we were able to enjoy a full hour of beekeeping anecdotes. Arthur was followed at Mountain Gray by John Eade who for 13 years continued to give beekeepers In the northern counties of England and beyond - an excellent service In beekeeping supplies. I look back with pleasure to my own visits to Holme-on-Spalding Moor over the years; and to the cups of tea which Mrs. Eade always managed to supply. It is good to find that Brian Eade is continuing to provide from his new premises at Westfield Avenue, Goole, the same friendly and efficient service which was the hallmark of his parents. I understand Brian and his wife operate some 300 stocks of bees. I wish this couple well.

A Visit to Mr Snelgrove

Snelgrove

From the early 1900's Mr Snelgrove's contributions to British and World bee-keeping have been immense and it was a great privilege to renew acquaintance with this remarkable personality. A past President and Hon. Life Member of the BBKA and author of three standard works relating to the craft - Mr Snelgrove, an octogenarian, retains that keen interest which has characterised the whole of his writings and his work. His book, "Queen Rearing" is in a class quite by itself.

Upon arrival at the Snelgrove home we sampled an excellent mead and then our host took us out among the bees. Over the years Mr Snelgrove has maintained his own strain of dark bees - as he finds them better suited to these islands - and particular attention has been paid to docility. Here, we witnessed a faultless demonstration of bee handling, slow deliberate movements and correct use of one's hands kept the bees under control. A number of mating boxes were being prepared to receive a late batch of queen cells - these were similar in many respects to the Swiss type I have used In Northumberland for experimental work.

Mr Snelgrove's extracting and honey storage house has been carefully planned and meticulous attention to detail was in evidence everywhere. A power-driven 20-comb Herrod-Hempsall Radial extractor is used and straining and storage tanks are set at convenient levels to facilitate the handling of honey. It was in this house that we noticed the wooden barrel, and subsequently learned one of the secrets of the excellent mead!

The storage shed for assembling hive parts and frames, etc., adjoins the honey house and this too, is well-appointed. This shed leads on to the garage so that it is a relatively easy matter to convey any piece of bee-keeping equipment to an out-apiary. The hive in use was a forerunner of the National and has been described in Mr Snelgrove's books. The floorboard is of special interest as the bees gain admission underneath. At the back, there is a large cavity which will take a special kind of feeder. In a bad summer, when there are supers on the hive, it is possible to feed the colony from below. Alexander, of Delanson in the USA, made use of a similar principle in his extensive apiary of 700 or more colonies.

In the evening Miss Snelgrove took us all by car to one of the out-apiaries, where the colonies are housed in a well-constructed bee house, erected by Mr Snelgrove himself a number of years ago. Here, the bees were storing well in the supers, and even in this disappointing summer, a pleasing crop of honey was in evidence.

We realised that afternoon that we had been guests of one of the most outstanding bee craftsmen Great Britain has produced.

When in the vicinity of Taunton we motored out to the lovely village of Lydeard St Lawrence to see our old friend Mr Leslie Hender, chairman of the Somerset B.K.A. Here, in charming surroundings Mr and Mrs Hender run, in a most successful manner, their colonies of bees. The Henders, too, showed us their bee house, which is situated in an Old World garden. This once belonged to John Spiller and is described in his book on bee houses as house No.2.

Donald Sims

Bee-keepers everywhere will be sorry to lean that my beekeeping associate, Donald Sims, has been laid low with illness during these past few weeks. Donald Sims entered northern beekeeping in 1963 with considerable beekeeping experience gained in Devon, Kent and Derbyshire, as well as other parts of England. A popular figure at beekeepers' gatherings all over the country, Sims, in a comparatively short tirne, has revitalised beekeeping thinking in the English Northern counties. A skillful practical beekeeper, Sims operates with Smith equipment and handles 30 stocks in Northumberland and 35 in Devon. As a young man Donald Sims knew the Sussex beemaster Samuel Simmins.

Messrs. Pope and Sims.

William Hamilton

A popular visitor to our apiary in the 1940's was William Hamilton, author of "The Art of Beekeeping". I have always enjoyed Hamilton's book and lectures. Like William Herrod-Hempsall, he could get an audience "going". It seems that lecturers of this kind have gone for good - and we are the poorer. When Hamilton was among the bees he rose to the occasion. In the summer of 1947 he tackled a powerful two brood chamber lot of yellow-banded Buckfast bees at Shilford.

Comb by comb the colony was examined but, according to Hamilton there was no trace of the queen. Again, after careful scouting, the combs were replaced. When Hamilton came to the outside comb - which had been removed from the hive at the beginning of the operation - for safety's sake - he exclaimed "she's here". A fact which I suspect he was aware of from the outset. The spectators had followed his examination in detail, and at this stage the excitement among the admirers was considerable.

Hamilton set out several systems of colony management in his book and excellent diagrams were provided by Mr. Lord. With heather honey in mind the system we liked best was this:

"Early in June, a nucleus should be made from each colony, or one or more colonies can be divided for the purpose. Ths nucleus should have two combs of sealed brood and enough bees cover four combs. A ripe queen cell should be given to each nucleus. Abundance of food and breeding room should be provided and, with careful attention, each nucleus should be a good colony on ten or more combs by the beginning of August. The temptation is strong to take these colonies as they stand on the moors. They should be united, however, one to each of the main colonies in the apiary. If desired, the old queens in the main colonies can be held in nuclei until the return from the moors when they can be killed, and on the whole this is good practice, as heather colonies usually need some help for winter.

The condition of each colony after uniting should be about ten combs of brood in each hive and enough bees to need at least twenty combs to hold them."

It would be wise to to implement this by providing combs of stores to the outside of the brood as a safeguard against bad weather.

Heather People: John and Rosemary Theobalds and the Robson Family of Horncliffe
(BBJ March 1989)

Cecil Tonsley with John Theobalds

Early in February we were delighted to welcome a well known Devon beekeeper, Len Davie, to Northumberland and Co. Durham. Devon born and bred Len has carried out bee disease inspection work with Roger Lancaster in Devon for many years.

In Northumberland John and Rosemary Theobalds entertained us at their home at Hexham. Older readers will remember that John was one of those who contributed "In the North" before it was taken over by the late Albert Hind for some years.

In the last 30 years John has become one of the most successful producers of heather comb honey in the country. He showed us some of his magnificent heather sections which he has taken off the Hexhamshire Fells. The late W.H-H., William Hamilton, Willie Smith of Innerleithen and Robson and Cessford of Northumberland would be proud of the fact that we have in John a man who is following faithfully In their footsteps. In the apiary John's colonies were in tremendous form and overwintering well with combs still packed with heather stores.

During Len Davie's stay we set out with Amy Nicholson and Basil and Winnie Waller to visit Florence and Selby Robson in North Northumberland and their family at Horncliffe nestling attractively above the River Tweed. On the way there the entire party managed to climb Ros Castle for its spectacular view over Chillingham Park. On some days it is possible to see the wild white cattle there.

Selby Robson

Now in their eighties, Florence and Selby are a remarkable couple and are an example of what can be achieved through endeavour and industry. Over the years with the aid of their son, William, and daughter-in-law, Daphne, they have built up a 1000 colony bee farm in the Scottish Border country with its impressive honey handling plant and equipment. Selby and his family turn out their own Smith hives with strong, properly-made floorboards. Hive floorboards generally made in this country are too flimsily made and our manufacturers should set out to correct this. Willie is full of drive and is wrapped up in the family business. He has sound ideas on marketing and the latest honey extractor was built locally.

Willie and Daphne Robson

As one of the best beemen in Europe, Selby Robson followed the teachings of the late Willie Smith of Innerleithen and Archie Ormiston of Stow, and acquired bees from beekeepers of their calibre over the years. Today he can proudly claim that the family bees are 'hefted' (like sheep are 'hefted' to a particular hill or moor) to the district that they the bees are in, and to which they really belong. I have visited the Robsons for 40 years now and it was a joy to listen to this great beekeeping family commenting on matters relevant to the 1990s and the approaching new Century.

Visit to Eva Crane at Hill House
(BBJ April, 1981)

Looking back to the year 1950, one can see that a whole era in British beekeeping was drawing to a close. Annie Betts climbed down from The Bee World editorial chair after stamping the paper with her unique style. William Herrod-Hempsall had been persuaded (I imagine by our present Editor, Cecil Tonsley) to come out of retirement and attend the National Honey Show: the great beekeeper was to die in September 1951 - the year the International Congress was held at Leamington Spa. Leonard Illingworth, for many years Secretary of the Apis Club also called it a day.

However, during the 1940s a new and dynamic figure had appeared on the beekeeping scene in Yorkshire, namely, Dr. Eva Crane. In July of that never-to-be-forgotten summer of 1949, when honey poured into the hive for weeks on end in most parts of the country, Dr. Crane visited Northumberland to speak at a large gathering of beekeepers at the apiary of the respected beekeeping partners, Messrs. Robson and Cessford of Riding Mill. Photographs taken at the event, which I have before me as I write, reveal a very young Dr. Crane holding an attentive audience spellbound. Now, thirty-two years later, I have been able to visit the headquarters of IBRA which Eva Crane and her helpers have built up in the succeeding years. In my own travels out of this country I have found that beekeepers are fascinated that this organisation is based here in the British Isles, and they are generally complimentary about the services IBRA provides.

On the afternoon of Tuesday, March 10th, Dr Crane and her team showed me their work at Hill House, Gerrards Cross. As Director of IBRA, Dr. Crane's own office is an imposing downstairs room which has a number of direct links with BBJ personalities of the past. The late Dr. Malcolm Fraser's bookcase is a superb collector's piece. Another interesting item is the C.N. Abbott clock. Charles Nash Abbott was the founder and first Editor of the BBJ. The IBRA team told me that Ted Hooper's book had become the best seller of our time and has taken the place of the Herrod-Hempsall, Cowan and Digges guide books of an earlier generation.

In Greece with Brother Adam
(BBJ March 1984)

Vlad Dermatopolous with Colin in Greece.

Early on January 12th I left Newcastle Airport for Heathrow to meet Brother Adam and we set out to revisit Greece. Brother Adam first visited this part of the Balkans In 1952 and his subsequent experiences with the native honey bee of the region - now known as Cecropia attracted worldwide attention. Angelo Komninos, whose father is an avid reader of the BBJ, met us off the plane at Athens and a happy evening and the following day was spent with this family. Later we met the Editor of the new Greek Bee Journal, Mr. Efstathios Vassillou, who publishes it in Athens.

Then we departed in glorious sunshine for the Peloponnese and the delightful old town of Nauplion, overshadowed by the Palamidi fortress and the castle of the Franks. Another beekeeping friend of longstanding, Tasios Tsakonas, greeted us at the Xenia Hotel, the views from which are lovely - the snow-capped rnountains surrounding the Argolic Gulf, along with the red-tiled roofs of the town below, and the Castello on an island in the harbour complete what is a charming picture.

Greek apiary at the Russian monastery of Agios Pendaelemonas.

The following day Tasios took us out among the bees and Brother Adam met and talked to the commercial beekeepers of the Peloponnese. After this we left for Northern Greece, Thessaloniki, and the Chalcidice peninsula. Another respected colleague of longstanding, Vlad Dermatopolous, with Dr. Thrasyvoulou and Panos Misirlis of the American Farm School welcomed us at the Airport at Salonika.

Since becoming a full member of the EEC, Agriculture in Greece has undergone an unbelievable transformation and this embraces beekeeping. Farming prospers and new combine harvesters and tractors are everywhere, along with modern buildings.

Soon we were on our way to Chalcidice, that strange three-pronged peninsula which has some of the most spectacular scenery in the whole of Greece. It was from Cassandra that Brother Adam obtained, in 1972, some of the finest Macedonian queens - and we returned to the Athos region of the peninsula in 1981 for the making of the Julian Pettifer film on Brother Adam's work.

The Beekeeping Department of the Chalcidice Agricultural Research Station has extensive apiaries in the region and steps are being taken with the aid of Instrumental Insemination to develop the best of the Macedonian lines of bees in Northern Greece. Dimitrios Tselios is in charge here, ably assisted by the botanist Maria Kostarelou-Damiouidou. Honeybees were working many of the plants in the trial grounds which she supervises and there were great sweeps of the following: Rosemary (Rosmarinus officinalis). Salvia (Salvia superba), Marjoram (Origanum heracleoticum). Marjoram (Origanum majorana), Thyme (Thymus serphyllum), Eucalyptus (meliotera). Bee Balm (Melissa officinalis), Thyme (Thymus capitatus), Lucerne (Medicago arborea) and Pittosporum (sinensis).

Three years ago the problem of the Varroa mite hung like a curse over beekeeping in Greece. Now, with the aid of Acaricides, particularly the powder Sinecar and the insecticide Malathion, the mite had been brought under control and is now no more of a problem than Braula infestation in the apiaries of progressive beekeepers. These beekeepers have again powerful colonies of bees and harvest large crops of surplus honey. The only Varroa mites that could be found were in the apiaries of the older bee men who do not believe in medicare and who rely entirely on culling out all the drone brood as control. In these apiaries all the drone pupae had three or four conspicuous and lively Varroa mites on them as the cells were opened.

As a sheep farmer this reminded me of our own problems 30 years ago - before the advent of prophylactic drugs. Losses among ewes and lambs were enormous but now, with the aid of vaccines we enjoy and maintain a wonderful control over all the known pests. The progressive Greek beekeepers have overcome their immediate problems in a similar way. Indiscriminate crop spraying from the air is at the moment a major problem for Greek beekeepers.

All too soon we left Thessaloniki for Crete, where another old friend, George Daskalakis who lives at Chania, awaited our arrival at the airport. The most southerly and largest of the Greek Islands, Crete enjoys days of warm sunshine throughout the winter, but the centre of the island is made up of great snow-clad mountains, mysterious and forbidding in appearance. George and Helena Daskalakis visited me in Northumberland in June, 1982, and the cool, bracing climate of the Border country was a revelation to them. George and Helena operate some 200 colonies of bees in double brood chamber Langstroth hives, which were all painted powder blue. Even in January a lot of pollen was being taken into the hives and the colonies had generally four fine combs of sealed brood each, George, too, has cleared the Varroa mite from his apiaries with the help of malathion. Every bee colony is dosed twice during the year, when supers are off the hives.

George claims that the native honey bee of Crete - which carries Brother Adam's name - is not unduly aggressive on the island. It is only when the searing heat from nearby Africa brings the honey flow abruptly to an end that the bees start to take it out on everyone. Those beekeepers who saw the Julian Pettifer film will recall the Cretan bees' mass attack on Brother Adam!

While on the island we called on the veteran beekeeper Christos Zimboulakis and his son who have a beekeeping museum at Akrotiri. Brother Adam met Christos on his first visit to Crete in 1952 and there is a picture of this celebrated bee man in "In Search of the Best Strains of Bees" holding a basket hive. The family supply pollen, propolis and royal jelly around the world. Newspaper cuttings on display indicate that our own Royal family are interested in these products of the hive.

We then enjoyed the hospitality of Jacob Tsouronakis and his German wife. This beekeeper's home has spectacular views over the charming old capital of Crete, Chania. A 9 pm flight from Hiraklion took us over the Aegean Sea on the final stages of our journey, to the well-known and beautiful island of Rhodes, where members of the Beekeepers' Co-operative met us. Jannis Alexakis, who had worked in Sweden for some years, acted as interpreter.

A variety of the Anatolian honey bee is found on the island and a different variety still on the island of Kos. Here, too, the Varroa mite has been brought under control and many of the bee colonies we were shown had 5 to 6 splendid Langstroth combs of sealed and hatching brood. Already in January there was a steady nectar flow. The Anatolian queens we spotted were literally the size of a part of the little finger and reminded me of the queen Brother Adam gave me in 1962 after his return from Turkey.

This heavily wooded and beautiful island is a beekeeping paradise and with the great tourist influx in summer the beekeepers are able to get more for their honey than we do in this country. The island's radlo announced Brother Adam's arrival on Rhodes and beekeepers from every district crowded into the Farm School to hear him speak.

Visit to Cyprus, 1979
(BBJ, November 1980)

My fascination for the part the honey bee of Cyprus has played when the bee-breeders of Italy and the U.S.A. set out to develop lines of light, almost transparent coloured, honey bees eventually lured me to the island. My visit coincided with the screening in this country of the BBC's TV spectacular "The Aphrodite Inheritance" and it was of interest to find that the Curium Palace Hotel, Limassol, which featured in the TV series, was also a meeting place for one or two of the island's bee-keepers.

Snow was lying deeply as I set out from Newcastle Airport on the morning of January 9th, 1979, but I was unaware that a winter destined to be as severe as 1947 and 1963 was in its early days. The incoming London plane left its flight path and floundered and it was some considerable time before we were given clearance.

The flight from Heathrow was subsequently delayed and we eventually arrived in Cyprus, cold, tired and miserable at midnlght. The transfer by car from Larnaca Airport into Limassol had still to be made. Warrn sunshine filled my room on the morning of January 10th and the discomforts of the previous day were soon forgotten. Roger White, a keen young bee-keeper, stationed at Akrotiri arrived at my hotel and took me by car to visit his apiaries on the island. Bees were flying strongly in the sun from his Langstroth hives and as Roger examined his colonies I was advised to stand well back in the undergrowth, as the spiteful occupants of the hives, zoomed round seeking something, or, someone to wreak their vengeance on. But, as a result of Roger's good handling and vigilance, no one was stung.

We then went on to visit the well-appointed apiary of Stellakis Panayides. Again, Langstroth hives were in use and were attractively spread out among the lemon trees. In this apiary the bees ignored us completely.

During my stay I visited Stavrovouni Monastery spectacularly situated on a mountain top midway between Limassol and Nicosia. There are apiaries everywhere in this area. It was to Stavrovouni that the enterprising Helena - mother of Constantine the Great, the Roman who set up the Christian faith we know today - brought a piece of the Cross on which Jesus was crucified which she had salvaged from the Holy Land. In this simple setting one of the great relics of Christianity is still preserved and the few monks offer hospitality to the devout believers who make the ascent to the mountain top. The views from the Monastery over the island are breathtaking.

From the warm sunshine of coastal Limassol and Paphos my bee-keeping friends look me to the snow bound heights of the Troodos Mountains where snowballing ard tobogganing were the order of the day. Then, there were further visits to well known bee-keepers on the island. The apiary at the Monastery of St. Barbara had old pipe hives as well as Langstroth hives. Apart from caring for the bees, the Mother Superior bred fish in large tanks in the monastery gardens. Again, in this apiary it was possible to go among the bees unprotected. In some apiaries the bees were docile, but in others the fiery disposition which has been attributed to the honey bees of Cyprus was quite pronounced.

St Ambrose of Milan, the Patron Saint of Beekeeping
BBJ St Ambrose

St Ambrose - patron Saint of Beekeepers -
a mosaic at the church of Saint Ambrogio, Milan

Over the years I have been approached for information about the Patron of Beekeepers, Ambrose of Milan, and I am always surprised that there is such scant reference to the activities of this Prelate in beekeeping books. On one of my visits to Milan I set out to find more about the beekeepers' friend.

Beekeepers who find themselves In Milan should visit Ambrose's own church known as Saint Ambrogio, an eleventh century building on the site of the church built by Ambrose and his followers in the fourth century. Inspect the panel in gold and silver at the high altar with its Ambrose motifs. The abbey scene depicts honey bees feeding the future leader of the Christian Church in Northern Italy, the implication being that this nurture resulted in Ambrose's splendid rhetoric and skilful use of words which helped to establish the Christian faith we know today, when Constantine the Great decided to adopt Christianity from among the many faiths then being practised. Also this provided human beings with a code of conduct and a firm structure to raise families. It provided countries of the then Empire with a labour force and soldiers for the many wars and made provision for the human need for spiritual nurture.

After a period of severe repression of all religion, Constantine issued the Edict of Milan in 313 AD which allowed Christians the same religious toleration as other faiths. For the next decade Milan, situated on one of the main trade routes in Europe, was in a constant state of turmoil over religious matters. The Ambrose family were of importance at Treves in Cisalpine Gaul, a region between the Alps and Appenines. Born In 340 AD, Ambrose was responsible, when he reached manhood, for maintaining law and order around Treves. Following his father's death Ambrose went to live in Rome with other members of the family. Here he studied and eventually practised Law. He then became Governor of Liguria in Northern Italy and it fell to him to establish law and order among the rival Christian factions of Milan.

At a time when a new Bishop of Milan was to be appointed, many of the early Christians regarded Jesus as a human figure and great missionary and evangeliser. To them, there was only one God - the Father and creator of all things. The rival group of Christians under the influence of Constantine and his mother Helen, had deified Christ at Nicaea in 325 AD and made him into a God as well, putting him on the same spiritual level as God the Father. From then on the early Christians were painstakingly eliminated and their beliefs and records destroyed.

After 380 AD only the Constantine beliefs were tolerated and accepted as the Christian faith. Ambrose's ability as a peacemaker among the warring Christian factions resulted in his being made Bishop of Milan in 374 AD. Up to this time he had not even been baptised. He was consecrated Bishop on December 7th, and this is known as St Ambrose's Day. For some time he struggled with Justlna, mother of the Western Roman Emperor, Valentinian II and champion of the early Christians, who wanted to share his church. It is said that he carried out a sit-in with his followers and shut the church doors in this formidable lady's face. During the sit-in Ambrose and his followers composed and sang DEUS CREATOR OMNIUM and TE DEUM for the first time.

Ambrose wrote many hymns and devised an arrangement of church music known as the Ambrosian Chant. At the time of his death, in 397 AD, Ambrose was one of the most influential men in the land and did his best to guide Constantine's wrangling successors. The present day custodians of the Sant Ambrogio will sometimes let visitors to the church see the good man's skeleton - if they are interested in such things.

APPENDIX

Comparative Costs of Hives (1961) in all cases without bees—

MODIFIED DADANT HIVE complete. Floorboard, Brood Chamber, Two Shallow Supers (frames fitted with foundation), Deep Roof ... £13 10s 6d

MODIFIED NATIONAL HIVE complete. Floorboard, Two Brood Chambers, Two Shallow Supers (frames fitted with foundation) Deep Roof ... £14 3s 0d

MODIFIED NATIONAL HIVE with one Brood Chamber, Three Shallow Supers (frames fitted with foundation), Deep Roof ... £13 10s 0d

SMITH HIVE complete. Floorboard, Two Brood Chambers, Two Shallow Supers (frames fitted with foundation), Deep Roof ... £10 9s 0d

NOTE: A Four Comb Nucleus Colony of bees should cost approximately £4.

ADVICE ON HANDLING BEES

One golden rule—never use gloves. Occasionally health and other reasons may demand it, but you can never acquire the deftness and quiet confidence which has characterised the great bee-masters of our time. When you open up your hives slow deliberate movements must continually be sought and all knocks and jars avoided. Remember that anyone who crushes bees during manipulation of hive parts, and refuses to improve his ways, doesn't know his job.

"Handle them like crusted port.
Never when your temper's short:
Lift the frames with tender care,
A sudden move will make you swear."

A.R.C. — Bee Craft.

THE BRITISH BEE-KEEPERS' ASSOCIATION

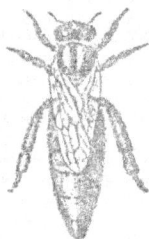

Coronation   Conference

1953

Organised by THE YORKSHIRE BEE-KEEPERS' ASSOCIATION

Patron: H.R.H. THE PRINCESS ROYAL

JUNE
26th Friday, 27th Saturday, 28th Sunday

AT THE

JOSEPH ROWNTREE

SECONDARY MODERN SCHOOL

NEW EARSWICK, YORK

By kind permission of the School Governors and the North Riding Education Committee

---

To be opened officially at 2 p.m., on June 27th, by

Brigadier-General Sir EDWARD WHITLEY, K.C.B.

Lecturers and Demonstrators

A. Abbott   T. Alton   Miss D. V. Burch, M.A.   C. G. Butler, M.A., Ph.D.

Mrs. R. E. Clarke   W. Dodd   A. Dines

G. W. Green, M.B.E., F.I.M.   W. Hamilton   L. Hay

A. Hebden, F.R.E.S.   E. Humphreys   F. Austin Hyde, M.A.

Miss M. Macfarlane   P. S. Milne, B.Sc.   A. V. Pavord

R. P. Sims   F. W. Smith   H. S. Thompson

C. C. Tonsley, F.R.E.S.   C. Weightman, F.R.E.S.

B.B.K.A.

President: D. S. HUDSON   *Chairman:* Dr. R. H. BARNES

Y.B.K.A.

President: THE VISCOUNTESS DOWNE   *Chairman:* T. R. FLYNN, M.C.

Y.B.K.A. Education Committee

Chairman: H. P. LOCKWOOD   *Hon. Sec.:* W. GEMMELL

Hon. Business Manager: A. R. MINNEY

Flyer for the Coronation Conference, 1953 -
the York even had a line-up of excellent speakers.

Northumberland Bee-keepers' Association.

MORPETH BRANCH.

....

- HONEY SHOW -

IN THE

Primary School, Castle Square, Morpeth

ON

SATURDAY, 11th OCTOBER, 1947.

Open to the Public from 2 to 7 p.m. Admission 6d.

....

23 CLASSES FOR HONEY AND BEESWAX.

(THE BIGGEST HONEY SHOW IN THE NORTH).

Judges : JAMES CUNNINGHAM, Esq. - Open & County Classes.

COLIN WEIGHTMAN, Esq.   - Branch Classes.

....

ADDITIONAL ATTRACTIONS :--

FILM SHOW : "Bee-keeping in the North."

APPLIANCES DISPLAY

by A. MILLER, Esq., Whittingham.

"TWENTY QUESTIONS" COMPETITION;

in charge of J. McNICOL, Esq., Editor, "Northern Apiarist."

Honey Judging Competition.   Bee-keeping Museum.

....

R. WOOD, Secretary.

Schedules from 32, North View, Newbiggin-by-Sea.

ENTRIES CLOSE OCT. 4th.

ARMSTRONG, PRINTER, HEXHAM

Northumberland 1947 Honey Show flyer -
"THE BIGGEST HONEY SHOW IN THE NORTH".

A group of beekeepers at Shilford. From left to right - Brian Ripley, Alnwick
(later BBKA Chairman), Basil Waller, George Batey
(President Newcastle BKA) & Trevor Green.

Bob Hawker at Shilford.

John Whent, a successful North Yorkshire
commercial beekeeper.

Great Yorkshire Show
Back Row 1. Bill Slinger 4. Harry Grainger Front Row 1. Mrs Neale 4. Mrs Ivy Jacques.

Great Yorkshire Show
(David Pearce, the Jeffersons and Michael Badger).

1955 BBKA Dinner, London.

Mr A. L. Hind of Washington, Durham.

Clive de Bruyn, CBI for Essex.

Roger Lancaster, Colin Weightman, Len Davie on Dartmoor when
visiting George Jenner's bees Sep: 1989

Walsingham Show 2010 - Colin with steward Kath Simon.

David Pearce who helps Colin with the bees at Shilford Jan: 2011

Jeff Rounce, from Norfolk, visiting Maurice McElrue's apiary at Blanchland, a moorland village in the Northern Pennines.

INDEX

ABBOTT: A. 91, 115

ABUSHADY: Dr. 64

ADAM: Bro. 12, 68, 78, 114

AMBROSE: 99, 128

APIARY HEALTH CERTIFICATE: 60

APIS CLUB: 63, 64, 66

BADGER: M. 135

BEST BEES: 92

BETTS: A. 8, 64, 121

BLANCHLAND: 29

BUCKFAST: 12

CARR: W.B. 65

CESSFORD: F. 11

CHARLTON: G. 92

COCKLE PARK: 32

COUSTON: R. 94

COOPER: Beo. 80

CRANE: Dr. Eva 70, 121

COWAN: T.W. 65

COX: S.J. 5, 33

CUNNINGHAM: J. 35, 71, 73, 100

DAVIES: J. 12

DONOVAN: P. 69

GAMBLE: R. 14

GLEED: J. 84

GRAY: H. 14

GREECE: 123, 124, 125

GREEN: G.W. 134

HAMILTON: W. 65, 117

HAWKER: R. 134

HERROD-HEMPSALL: W. 65, 66, 67

HOUGHALL: 5

HUMPHREYS: T. 34, 100, 106

HYDE: A. 7, 76

HUNTER: A. 17

ILLINGWORTH: L. 63, 111, 121

JEFFERSON: A. 135

JOBSON: G. 7

JORDAN: J. 15

KHALIFMAN: I.A. 53, 101

KERR: R. 11

LOGAN: M. 72

LIMOND: A. 71

MORISON: Dr. G. 94

MANLEY: R.O.B. 92, 93, 100

MILNE: P.S. 60

NEWTON RIGG: 13

NOSEMA: 60, 92

PAWSON: C. 7, 31

PEARCE: D. 138

PHIPPS: J. 3, 89

ROUNCE: J. 139

SECKSTONE: T. 60, 66

SKILLING: R. 73

SLINGER: W. 65, 135

SNELGROVE: L.E. 13, 113, 115

SIMS: D. 100, 117

SIMMINS: S. 59, 117

STEEL & BROODIE: 97, 99

ROBSON: R. 24

ROBSON: D. 121

ROBSON: S. 119, 120

ROBSON: W. 100, 121

THEOBALD: J & R. 119

THORNE: E. H. 97

WADEY: J. 74

WALLER: B. 134

WEIGHTMAN: W & H. 17

WHENT: J. 91, 134

WHITE: R. 127

www.ingramcontent.com/pod-product-compliance
Lightning Source LLC
Chambersburg PA
CBHW051337200326

41519CB00026B/7453